T0192367

Communications in Computer and Information Science 2057

Rationale

The CCIS series is devoted to the publication of proceedings of computer science conferences. Its aim is to efficiently disseminate original research results in informatics in printed and electronic form. While the focus is on publication of peer-reviewed full papers presenting mature work, inclusion of reviewed short papers reporting on work in progress is welcome, too. Besides globally relevant meetings with internationally representative program committees guaranteeing a strict peer-reviewing and paper selection process, conferences run by societies or of high regional or national relevance are also considered for publication.

Topics

The topical scope of CCIS spans the entire spectrum of informatics ranging from foundational topics in the theory of computing to information and communications science and technology and a broad variety of interdisciplinary application fields.

Information for Volume Editors and Authors

Publication in CCIS is free of charge. No royalties are paid, however, we offer registered conference participants temporary free access to the online version of the conference proceedings on SpringerLink (http://link.springer.com) by means of an http referrer from the conference website and/or a number of complimentary printed copies, as specified in the official acceptance email of the event.

CCIS proceedings can be published in time for distribution at conferences or as post-proceedings, and delivered in the form of printed books and/or electronically as USBs and/or e-content licenses for accessing proceedings at SpringerLink. Furthermore, CCIS proceedings are included in the CCIS electronic book series hosted in the SpringerLink digital library at http://link.springer.com/bookseries/7899. Conferences publishing in CCIS are allowed to use Online Conference Service (OCS) for managing the whole proceedings lifecycle (from submission and reviewing to preparing for publication) free of charge.

Publication process

The language of publication is exclusively English. Authors publishing in CCIS have to sign the Springer CCIS copyright transfer form, however, they are free to use their material published in CCIS for substantially changed, more elaborate subsequent publications elsewhere. For the preparation of the camera-ready papers/files, authors have to strictly adhere to the Springer CCIS Authors' Instructions and are strongly encouraged to use the CCIS LaTeX style files or templates.

Abstracting/Indexing

CCIS is abstracted/indexed in DBLP, Google Scholar, EI-Compendex, Mathematical Reviews, SCImago, Scopus. CCIS volumes are also submitted for the inclusion in ISI Proceedings.

How to start

To start the evaluation of your proposal for inclusion in the CCIS series, please send an e-mail to ccis@springer.com.

Quan Yu

Editor

Space Information Networks

7th International Conference, SINC 2023
Wuhan, China, October 12–13, 2023
Revised Selected Papers

 Springer

Editor
Quan Yu
Institute of China Electronic Equipment System
Engineering Corporation
Beijing, China

ISSN 1865-0929   ISSN 1865-0937 (electronic)
Communications in Computer and Information Science
ISBN 978-981-97-1567-1   ISBN 978-981-97-1568-8 (eBook)
https://doi.org/10.1007/978-981-97-1568-8

This Springer imprint is published by the registered company Springer Nature Singapore Pte Ltd.
The registered company address is: 152 Beach Road, #21-01/04 Gateway East, Singapore 189721, Singapore

Paper in this product is recyclable.

Preface

This book collects the papers presented at the 7th Space Information Network Conference (SINC 2023), an annual conference organized by the Department of Information Science, National Natural Science Foundation of China. SINC is supported by the key research project of the basic theory and key technology of space information network of the National Natural Science Foundation of China, and organized by the "space information network" major research program guidance group. Its aim is to explore progress and new developments in space information networks and related fields, to show the latest technology and academic achievements in space information networks, to build an academic exchange platform for researchers in China and abroad working on space information networks and in related industry sectors, to share achievements and experience in research and applications, and to discuss new theories and new technologies of space information networks.

This year, SINC received 73 submissions, including 43 English papers and 30 Chinese papers. After a thorough reviewing process in which papers received on average three double-blind reviews, 13 outstanding English papers were selected for this volume (retrieved by EI), accounting for 30.2% of the total number of English papers.

The high-quality program would not have been possible without the authors who chose SINC 2023 as a venue for their publications. We are also very grateful to the Academic Committee and Organizing Committee members, who put a tremendous amount of effort into soliciting and selecting research papers with a balance of high quality, new ideas, and new applications.

We hope that you enjoy reading and benefit from the proceedings of SINC 2023.

December 2023   Quan Yu

Organization

SINC 2023 was organized by the Department of Information Science, National Natural Science Foundation of China; the Department of Information and Electronic Engineering, Chinese Academy of Engineering; China InfoCom Media Group; and the *Journal of Communications and Information Networks*.

Organizing Committee

General Chairs

Quan Yu	Institute of China Electronic Equipment System Engineering Corporation, China
Jianya Gong	Wuhan University, China
Jianhua Lu	Tsinghua University, China

Steering Committee

Zhixin Zhou	Beijing Institute of Remote Sensing Information, China
Hsiao-Hwa Chen	National Cheng Kung University, Taiwan, China
George K. Karagiannidis	Aristotle University of Thessaloniki, Greece
Xiaohu You	Southeast University, China
Dongjin Wang	University of Science and Technology of China, China
Jun Zhang	Beihang University, China
Haitao Wu	Chinese Academy of Sciences, China
Jianwei Liu	Beihang University, China
Zhaotian Zhang	National Nature Science Foundation of China, China
Xiaoyun Xiong	National Nature Science Foundation of China, China
Zhaohui Son	National Nature Science Foundation of China, China
Ning Ge	Tsinghua University, China
Feng Liu	Beihang University, China

Mi Wang   Wuhan University, China
Chang Wen Chen   State University of New York at Buffalo, USA
Ronghong Jin   Shanghai Jiao Tong University, China

Technical Program Committee

Jian Yan   Tsinghua University, China
Min Sheng   Xidian University, China
Junfeng Wang   Sichuan University, China
Depeng Jin   Tsinghua University, China
Hongyan Li   Xidian University, China
Qinyu Zhang   Harbin Institute of Technology, China
Qingyang Song   Northeastern University, China
Lixiang Liu   Chinese Academy of Sciences, China
Weidong Wang   Beijing University of Posts and
                                    Telecommunications, China
Chundong She   Beijing University of Posts and
                                    Telecommunications, China
Zhihua Yang   Harbin Institute of Technology, China
Minjian Zhao   Zhejiang University, China
Yong Ren   Tsinghua University, China
Yingkui Gong   University of Chinese Academy of Sciences,
                                    China
Xianbin Cao   Beihang University, China
Chengsheng Pan   Dalian University, China
Shuyuan Yang   Xidian University, China
Xiaoming Tao   Tsinghua University, China

Organizing Committee

Chunhong Pan   Chinese Academy of Sciences, China
Yafeng Zhan   Tsinghua University, China
Liuguo Yin   Tsinghua University, China
Jinho Choi   Gwangju Institute of Science and Technology,
                                    South Korea
Yuguang Fang   University of Florida, USA
Lajos Hanzo   University of Southampton, UK
Jianhua He   Aston University, UK
Y. Thomas Hou   Virginia Polytechnic Institute and State
                                    University, USA

Ahmed Kamal	Iowa State University, USA
Nei Kato	Tohoku University, Japan
Geoffrey Ye Li	Georgia Institute of Technology, USA
Jiandong Li	Xidian University, China
Shaoqian Li	University of Electronic Science and Technology of China, China
Jianfeng Ma	Xidian University, China
Xiao Ma	Sun Yat-sen University, China
Shiwen Mao	Auburn University, USA
Luoming Meng	Beijing University of Posts and Telecommunications, China
Joseph Mitola	Stevens Institute of Technology, USA
Sherman Shen	University of Waterloo, Canada
Zhongxiang Shen	Nanyang Technological University, Singapore
William Shieh	University of Melbourne, Australia
Meixia Tao	Shanghai Jiao Tong University, China
Xinbing Wang	Shanghai Jiao Tong University, China
Feng Wu	University of Science and Technology of China, China
Jianping Wu	Tsinghua University, China
Xiang-Gen Xia	University of Delaware, USA
Hongke Zhang	Beijing Jiaotong University, China
Youping Zhao	Beijing Jiaotong University, China
Hongbo Zhu	Nanjing University of Posts and Telecommunications, China
Weiping Zhu	Concordia University, Canada
Lin Bai	Beihang University, China
Shaohua Yu	FiberHome Technologies Group, China
Honggang Zhang	Zhejiang University, China
Shaoqiu Xiao	University of Electronic Science and Technology of China, China

Contents

A System-Level Entity-in-the-Loop Simulation Platform for Space-Terrestrial Integrated Network

Houze Liang[✉], Xinhua Zheng, Junrong Li, Tianyu Lu, and Xiang Chen

Sun Yat-sen University, Guangzhou, China
lianghz3@mail2.sysu.edu.cn

Abstract. As terrestrial mobile networks continue to converge with Low Earth Orbit (LEO) satellite systems to augment connectivity and facilitate low-latency services, the imperative for exhaustive validation mechanisms grows increasingly salient. This paper introduces a flexible space-terrestrial integrated testbed designed to address this gap. The testbed enables the flexible validation of diverse network architectures, focusing on systemic challenges in satellite networks. The paper further analyzes the exacerbating impact of high mobility and handovers on long-haul signaling procedures in space-terrestrial integrated networks. Utilizing systemic simulation and a satellite-based edge core network deployment strategy facilitated by the testbed, we achieve improvements in access robustness. Lastly, the testbed's adaptability and potential as a tool for future network design and optimization are also highlighted.

Keywords: Space-Terrestrial Integrated Communication · LEO Satellites · Entity-in-the-loop (EIL) Testbed

1 Introduction

Over the past few decades, mobile communication networks have improved a lot, moving from 2G to 5G systems. These improvements have increased speed and bandwidth and have also made services available to billions of people around the world. However, there are still challenges like high costs and limited coverage, especially in remote and sea areas.

To mitigate the coverage gaps inherent in conventional mobile networks in these specialized settings, LEO satellite systems have become a popular solution. Initiatives like Iridium and Globalstar pioneered the use of LEO satellites for global communications as early as the 1990s. These early efforts faced problems with cost and technology. With the advent of the 21st century, advances in technology and better business plans have made LEO satellite systems more promising. Notably, projects such as SpaceX's Starlink [1] and OneWeb [2] have improved these systems by offering faster data speeds and lower delays while achieving near-global coverage via extensive satellite constellations.

As work on Beyond 5G (B5G) and 6G technologies continues, there's a growing effort to combine these new mobile technologies with LEO satellite networks [3]. The

© The Author(s), under exclusive license to Springer Nature Singapore Pte Ltd. 2024
Q. Yu (Ed.): SINC 2023, CCIS 2057, pp. 1–8, 2024.
https://doi.org/10.1007/978-981-97-1568-8_1

goal is to provide high-quality, everywhere access to meet the changing communication needs of the future. However, this integration introduces complexities, including costly capital expenditures (CapEx) and challenges in comprehensive validation.

Current simulation tools for validation, such as OPNET for satellite network [4, 5] and UERANSIM/free5GC [6, 7] for mobile networks, each offer unique advantages but also limitations. The former excels in managing complex network topologies but is not as good at simulating detailed behaviors of individual network node behaviors. Conversely, node-wise protocol simulations tools that focus on these details struggle with bigger, system-level challenges.

A prominent issue is that relying solely on any single tool makes it difficult to capture various complex factors in space-terrestrial integrated systems, such as congestion, signaling storms, and other special scenarios like high-speed satellite moving and dense access caused by urban tides. Study [8] employs system-level simulation modeling at the PHY and MAC layers, providing a detailed description of high mobility and high downlink interference switching scenarios in LEO satellite networking, thereby analyzing the performance of switching algorithms. This method allows for customized analysis and optimization of the handover process but is limited in its applicability to large-scale issues such as network architecture improvements. Study [9] addresses the issue of signaling storms generated by state migration during high-speed satellite transits and uses business dataset [10] to model this situation, analyzing and validating the architectural improvements proposed in the article. However, due to the general shortage of business datasets for space-terrestrial integrated networks, it is difficult to validate updated procedures when the network architecture is updated. This makes traditional simulation approaches inadequate for comprehensively and accurately describing and optimizing specific scenarios.

In order to tackle the aforementioned problem, this paper integrates the aforementioned large-scale network simulation with detailed protocol stack simulation tools, while putting forward a system-level simulation method based on the EIL paradigm. The aim is to provide a more comprehensive and accurate simulation of the performance of LEO satellite systems. Employing this methodology enables us to leverage actual node-generated business datasets while facilitating nuanced analyses of specialized scenarios, including control plane signaling congestion and extended latencies.

The remainder of this paper is organized as follows. Section 2 discusses the features and architecture of the space-terrestrial integrated EIL simulation platform, presents the deployment of satellite edge AMF/SMF/UPF network topology based on its systematic feature. Section 3 conducts experiments on the aforementioned network topology, analyzes the system latency to user access both in normal and congestion scenarios. This section demonstrates the system's capability to analyze metrics in specific architecture workflows in space-terrestrial integrated networks. Section 4 concludes the paper and outlines avenues for future research.

2 Air Ground Integrated System-Level Platform

The space-terrestrial integrated EIL topology is composed of two main components: the EIL simulation platform and the customized deployment utilizing this platform.

The logical architecture of the simulation platform, as depicted in Fig. 1, demonstrates the interconnection of system components and the workflow process. To accurately depict specific scenarios and workflows in space-terrestrial integrated networking, such as signaling storms during handovers and hotspot area access, we introduce the EIL simulation platform. This platform employs a nearly complete protocol stack to simulate real interactions within the network and replicate dedicated scenarios. Based on this simulation platform, diverse B5G satellite core network topologies can be constructed to simulate the aforementioned customized scenarios.

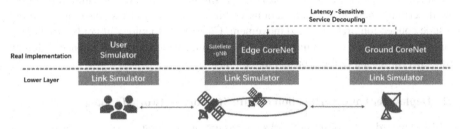

Fig. 1. Lower and upper layer modeling of nodes

2.1 Space Terrestrial Integrated EIL Simulation Platform

The EIL simulation platform consists of EIL nodes and a simulation medium environment. The former includes EIL modeling of satellites, ground stations, and users, while the latter describes the interaction or medium between EIL nodes, including their positional and visibility relationships, and the status of connected channels.

EIL Modeling of Integrated Nodes. EIL simulation is employed to precisely emulate procedures or workflows within specified architectures, activated by targeted signaling behaviors. The key to this approach is the EIL modeling of satellites, ground stations, and users, which we call integrated nodes. Each node's EIL model is split into two main parts: the basic infrastructure and lower-layer communication layer, and the higher-level protocols that connect to the real nodes. This two-part model gives us a detailed look at how each node actually behaves.

The core infrastructure mainly uses statistical models to simplify real network connections. Our simulation platform leverages OPNET's packet mechanisms to manage the complexities of lower layers, ensuring reliable point-to-point data transfer. For aspects like retransmission requests, we've employed statistical models, such as bit error rate and packet loss rate. This approach is sufficient for our goal: to simulate specific behavior in given architecture like control and user-plane congestion across various network setups, without adding unnecessary complexity.

The upper protocol layer defines the capabilities and responses of simulated nodes. We've customized open-source 5G core network software free5GC [6] and 5G access network software UERANSIM [7] to fit the interaction logic in our integrated EIL simulation platform. These customizations are integrated into the core infrastructure, allowing for consistent interactions within the simulated network. To enable flexible setup of base stations and core network components, we've modularized the protocol stack software. This modularity allows for the distributed deployment of latency-sensitive functions in AMF/SMF/UPF, either on satellites or other edge locations, thereby enhancing the platform's adaptability for testing various network architectures.

EIL Node Interaction Simulation. In the construction of the EIL simulation platform, EIL modeling of node interactions occupies a central position. Based on the afore-mentioned modeling, the system provides detailed depictions of the interactions and intermediate steps between these basic units, simulating complex dynamic topologies and describing these nodes' interactions within the topology. These basic interactions include connection, tracking, and movement. Changes in topology affect basic interactions like connection, communication, and tracking, which in turn influence different strategies composed of these basic interactions.

2.2 Deployment and Verification of Satellite Core Network Topology

Unlike standard 5G mobile networks, space-ground integrated networks face unique challenges like on-board routing and managing ground user access. We've adapted the AMF/SMF/UPF functions from 5G to work on satellite nodes and introduced new algorithms tailored for the fast-changing conditions in LEO satellite networks.

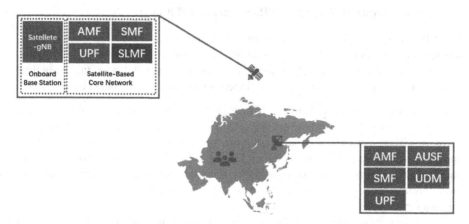

Fig. 2. Deployment topology of satellite core network

Figure 2 shows that LEO satellites serve as base stations, allowing direct links with ground users. Other core network functions, like UDM/AUSF and less time-sensitive AMF/SMF/UPF tasks, are handled at ground stations. Ground users connect directly to these on-board base stations, and both satellite and ground networks work together to manage user data and control signals.

We've also added a Satellite Link Management Function (SLMF) to the system, which is unique to space-based networks. SLMF helps collect and store global link data and operates under various strategies. Figure 3 shows how it monitors and manages:

(1) The connectivity between adjacent satellites
(2) The quality of these inter-satellite links
(3) Global inter-satellite link information of the entire LEO network

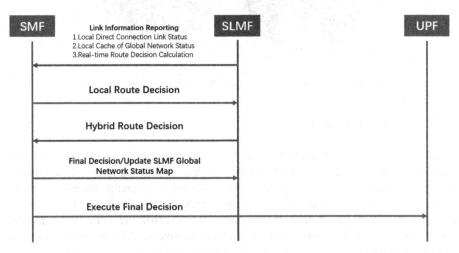

Fig. 3. SLMF decision procedure

SLMF with the previously mentioned state data plays an active role in network routing. It alleviates the routing complexity arising from the fast-changing and periodic nature of LEO satellites by utilizing comprehensive inter-satellite link data. For routing decisions, SLMF relies on its local topology data and periodically sends key information to SMF, allowing for more informed routing choices.

The configuration of the LEO constellation consists of 12 * 18 satellites forming a rose constellation, with its constellation parameters shown in Table 1 and Fig. 4. There are a total of seven ground stations, mainly located in mainland China, and users are deployed globally according to a configured distribution.

Table 1. Parameters of the 216-satellite inclined orbit rose constellation

Orbital Altitude	Orbital Inclination	Total Number of Satellites	Number of Orbital Planes	Number of Satellites per Orbital Plane	Phase Factor
1150 km	50°	216	12	18	1

Control of direct satellite-to-earth links and on-orbit routing is implemented at upper-layer signaling procedure handling functions. The SLMF component operates as a strategic plugin within the on-board protocol stack, linked to a specialized inter-satellite link management module. By aggregating link data for routing decisions, it contributes to control-plane routing led by SMF/UPF.

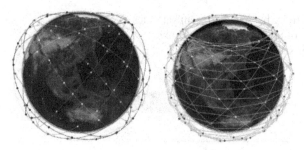

Fig. 4. 216 rose constellation and inter-satellite link connectivity

3   Experimental Results

Based on the topology delineated in the preceding section, we executed a series of experiments to evaluate critical performance metrics, including latency and congestion, throughout the access procedure. Beyond the topology and parameters outlined in Sect. 2.2, we exploited the discrete-event simulation capabilities of our system to configure the processing delay to be substantially lower than the associated transmission delay. The ensuing results and their comprehensive analysis are detailed in the subsequent subsections.

3.1   Satellite Access Latency

Figure 5 contrasts the latency during the access registration procedure between the architecture proposed herein and the "No Edge" topology, where the satellite serves only as an access base station and forwarding relay interacting with users and the ground core network without edge NFs. The edge-integrated terrestrial-satellite framework demonstrates a marked reduction in latency during onboard satellite access. This design leverages the locational advantages of edge deployment, significantly diminishing transmission time over N2 interfaces and curtailing the latency induced by multi-hop satellite links. In scenarios involving extended multi-hop transmissions, both round-trip times and computational delays adversely affect satellite mobility. During simulations, there is a likelihood that the authentication procedure remains incomplete; this, coupled with satellite mobility, necessitates intra-satellite beam switching, thereby triggering an expedited reconnection or even necessitating a re-initiation of the authentication procedure.

Figure 6 illustrates the probability of either reconnection or intra-satellite beam switching during protracted procedures for both traditional and proposed topologies. Owing to the inherent multi-hop nature of these extended procedures, the latency incurred frequently surpasses that of terrestrial mobile communication systems. This results in an increased likelihood of enforced reconnections or switching, exacerbated by the high-velocity moving of satellites and on-orbit congestion. The strategic placement of latency-sensitive network service at the access satellite can mitigate these challenges by eliminating superfluous inter-satellite link relay steps, thereby not only reducing transmission latency but also alleviating the uncertainties introduced by prolonged congestion.

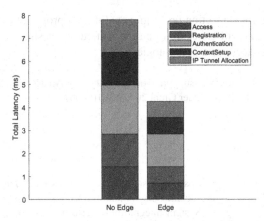

Fig. 5. Comparative latency analysis across various procedures in edge-deployed AMF/SMF/UPF architecture

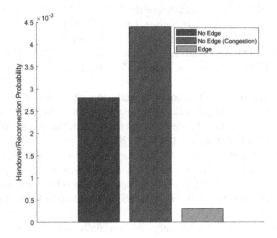

Fig. 6. Probability of reconnection/intra-satellite beam switching during extended procedures across different deployment strategies, averaging 500 users per satellite

4 Conclusion

This research presents the formulation of an EIL simulation platform tailored for space-terrestrial integrated systems. In light of the complexities associated with extended deployment cycles and CapEx in next-generation integrated space-terrestrial mobile communication ecosystems, our methodology leverages the software-defined attributes intrinsic to various networks. By employing system-level EIL simulations, we furnish a versatile framework that serves dual purposes: the design and validation of network architectures. As an illustrative case, we scrutinize the congestion risks inherent in satellite access and handover/reconnection procedures within our proposed network configuration. Future endeavors will concentrate on the meticulous design and optimization of diverse network architectures and topologies.

Acknowledgments. This work was supported in part by the Key-Area Research and Development Program of Guangdong Province under Grant 2019B010158001.

References

1. SpaceX Starlink Homepage. https://www.starlink.com/. Accessed 20 Jan 2024
2. OneWeb constellation Homepage. https://www.oneweb.world/. Accessed 20 Jan 2024
3. Chen, S., Sun, S., Kang, S.: System integration of terrestrial mobile communication and satellite communication—the trends, challenges and key technologies in B5G and 6G. China Commun. **17**(12), 156–171 (2020)
4. Jia, Y., Peng, Z.: The analysis and simulation of communication network in Iridium system based on OPNET. In: Proceedings of the 2010 2nd IEEE International Conference on Information Management and Engineering, Chengdu, China, pp. 68–72 (2010)
5. Wang, P., Zhang, J., Zhang, X., et al.: Performance evaluation of double-edge satellite terrestrial networks on OPNET platform. In: Proceedings of the 2018 IEEE/CIC International Conference on Communications in China (ICCC Workshops), Beijing, China, pp. 37–42 (2018)
6. Free5gc Homepage. https://github.com/free5gc/free5gc. Accessed 20 Jan 2024
7. UERANSIM Homepage. https://github.com/aligungr/UERANSIM. Accessed 20 Jan 2024
8. Kim, J., Lee, J., Ko, H., et al.: Space mobile networks: satellite as core and access networks for B5G. IEEE Commun. Mag. **60**(4), 58–64 (2022)
9. Li, Y., Li, H., Liu, W., et al.: A case for stateless mobile core network functions in space. In: Proceedings of the ACM SIGCOMM 2022 Conference, New York, NY, pp. 298–313 (2022)
10. Juan, E., Lauridsen, M., Wigard, J., et al.: 5G new radio mobility performance in LEO-based non-terrestrial networks. In: 2020 IEEE Globecom Workshops (GC Wkshps), pp. 1–6. IEEE (2020)
11. SpaceCore-SIGCOMM22.   https://github.com/yuanjieli/SpaceCore-SIGCOMM22. Accessed 20 Jan 2024

A Content-Based Generator Method for Vessel Detection

Yang Yang[1], Zhenzhen Xu[1], Xingyu Liu[1], Jun Pan[1(✉)], and Likun Liu[2]

[1] State Key Laboratory of Information Engineering in Surveying, Mapping and Remote Sensing, Wuhan University, 129 Luoyu Road, Wuhan 430079, China
panjun1215@whu.edu.com

[2] China Ship Development and Design Center, 268 Zhangzhidong Road, Wuhan 430064, China

Abstract. With the advancement of satellite imaging technology, the interpretation of remote sensing (RS) images has become an important subject. Especially in the object recognition field, accurate acquisition of vessel location and classification information is crucial to develop strategic plans. However, the lack of ship samples in RS images has been hindering the research of ship fine categorization. In this paper, a vessel generation method based on generative adversarial network is proposed to solve the insufficient samples in RS images. Dealing with sample insufficiency by prior global-local segmentation, category image generation and domain translation composition. Experiments on the HRSC2016 dataset show that the generated pseudo-images are highly similar to the real vessel, which verifies the effectiveness of the method. Besides, we constructed a ship dataset containing 10,000 images, which have great significance in vessel classification and localization.

Keywords: Data Expansion · Deep Learning · Vessel Detection · Contextual Separation

1 Introduction

Remote sensing (RS) image analysis has always been a significant research focus in the fields of computer vision and machine learning. It finds crucial applications in various domains such as maritime traffic surveillance, environmental monitoring, and marine management [1–3]. In recent years, with advancements in satellite imaging technology and the availability of high-resolution RS data, efficient interpretation of RS images has become increasingly important.

Particularly in the safety and civil domain [4–6], obtaining precise vessel location and classification information is crucial for detecting the movement trajectories of vessels, enabling strategic planning, and ultimately achieving accurate targeting of ships. An excellent ship classification and recognition algorithm must be capable of correctly detecting instances belonging to the ship category, identifying the specific subcategory of the object, and accurately locating the detected ship [7, 8]. However, most previous research has primarily focused on distinguishing between ships and backgrounds or a

Q. Yu (Ed.): SINC 2023, CCIS 2057, pp. 9–16, 2024.
https://doi.org/10.1007/978-981-97-1568-8_2

few broad categories such as commercial and military vessels. There has been limited research on fine-grained ship classification in RS images. Due to the limited availability of annotated ship datasets and the small inter-class differences in terms of shape, color, and texture among ships, conventional object detection and recognition methods struggle to achieve precise and fine-grained ship classification and localization [9, 10].

For fine-grained image classification and localization tasks, there exist well-known large-scale natural datasets such as CUB-200-2011 [11], Stanford Cars, and Aircraft datasets that have significantly contributed to the advancement of fine-grained image classification and localization tasks in natural scenes. Although features learned from natural images through deep convolutional networks can be fine-tuned to transfer to RS images [12], the differences in data distribution between natural and RS images cannot be eliminated. This causes models fine-tuned for natural images to perform poorly on fine-grained ship classification and location tasks for RS images. Despite the progress made using deep learning methods in object detection and recognition in RS images, the lack of a publicly available large-scale fine-grained ship classification dataset in RS images continues to pose a challenge for fine-grained ship classification and localization.

To accelerate research in fine-grained ship classification and localization in optical RS images, this paper proposes a ship dataset augmentation method based on Generative Adversarial Networks (GANs) [13] and utilizes this method to create a subclass-rich ship classification and localization dataset. We collected RS images from Google Earth and popular RS image datasets like HRSC2016 [14], covering various class scales of ships. The dataset comprises 10,000 images across four categories, encompassing different types of ships, various lighting conditions, and background environments. Additionally, compared to the ship data in the HRSC2016 dataset previously used for ship classification, the number of instances for each ship category has significantly increased. This greatly enhances the task of fine-grained ship classification and localization in RS images. Our primary contributions in this work are as follows:

1. We propose a novel ship data augmentation method that can expand ship samples to approximate real images, even with limited samples. This method can provide a good data base for ship detection.
2. We create a diverse RS image fine-grained ship classification and localization dataset (HRSC-warcraft10k), which consisting 10,000 images

2   The Proposed Method

2.1   Prior Global-Local Segmentation

In ship training background images, objects resembling ship features, such as bridges, dams, buildings, and containers, often appear. Unfortunately, this presence can be detrimental to the learning and training of deep convolutional neural networks. To address this issue, it is essential to implement prior global-local segmentation, aiming to eliminate the disruptive influence of complex background information. The primary processes involved in achieving this goal encompass mask extraction, straightforward data enhancement, data augmentation-based domain transfer [15], and pseudo-image generation.

By effectively distinguishing between the foreground and background, we can significantly reduce the noise impact in ship images [16]. This segmentation process yields cleaner images, leading to improved accuracy and performance when undertaking downstream tasks such as image analysis, feature extraction, target detection, and image classification.

2.2 Category Image Generation

The ships conform to a standardized design, characterized by sleek and linear aesthetics. Therefore, by implementing style normalization based on these consistent attributes, a broader spectrum of ship samples can be generated.

The generator in StyleGAN2 is a layered network, wherein each layer is responsible for generating distinct details within the image. Initially, a Mapping network transforms the input random noise vector into an intermediate representation vector. Subsequently, a style modulator integrates style information with the intermediate representation vector to govern the appearance and characteristics of the generated image at every level of generation. Finally, these features undergo convolution to produce specific details and textures.

The discriminator part of StyleGAN2 adopts a progressive growing strategy, commencing with the generation and evaluation of images at lower resolutions and gradually advancing to generate higher-resolution images. This process incorporates normalization layers as well as instance normalization layers. Moreover, StyleGAN2 employs equalized learning rates to ensure equitable updates for each layer's weights. Lastly, it estimates the standard deviation for each mini-batch in order to evaluate the quality of the generated image.

2.3 Domain Translation Composition

In the context of training a Generative Adversarial Network (GAN) specifically tailored for object detection tasks, we have meticulously devised a sequential approach to generate images that encompass background information while ensuring precise annotation of ship objects in terms of their position and size. The following section outlines the methodology employed:

1) Local Sample Instance Generation

The primary objective of our study was to eliminate any potential interference from the background and concentrate solely on generating high-quality images of the ship. This approach enables Gans to effectively learn and capture the visual features as well as morphology of ships, resulting in a consistent and diverse set of samples that style and type.

2) Global Synthesis based on Color Consistency

To generate a ship-free background image, we employed a mask to accurately delineate the boundary between water and land areas in the raw data. Subsequently, we assigned a specific pixel value to represent the water region and uniformly applied it to all pixels within this area. This processing step yields an image processed ship-free water region. By employing pixel-level background replacement, we successfully obtain a pristine and impurity-free sea surface background image, which serves as an ideal backdrop for synthesizing pseudo-samples.

3) Overall Scene Details Composition

Ultimately, the ship objects generated were randomly placed within the water body regions of the pure background images. This randomization process was introduced to replicate the diversity observed in real-world scenarios, thereby avoiding the imposition of excessive regularity on the data that could potentially compromise the generalization performance of the object detection model.

Through this sequence of steps, composite RS images that incorporate background information were successfully created. These images provide additional contextual information and diversify object detection tasks. This methodology significantly contributes to enhancing robustness, facilitating adaptation to various backgrounds and scenes, and ultimately resulting in exceptional performance in practical applications (Fig. 1).

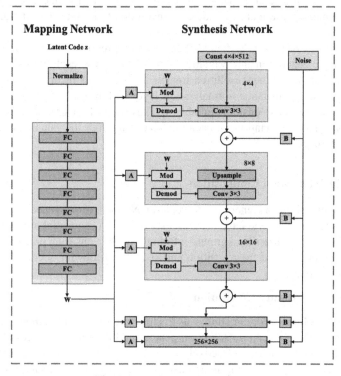

Fig. 1. StyleGANv2 framework.

3 Experimental Evaluation

3.1 Dataset Description

The data used in this experiment is from the HRSC2016 dataset, which was released by Northwestern Polytechnical University in 2016. It includes images from two scenarios: open-sea vessels and nearshore vessels. All the images were collected from six well-known ports, encompassing default imagery from Google Earth as well as corresponding historical data. The images in this dataset are annotated in the Oriented Bounding Box (OBB) format, with resolutions ranging from 2 m to 0.4 m and image sizes varying from 300 × 300 to 1500 × 900, with most exceeding 1000 × 600. In total, there are 1061 images with valid annotations distributed across training, validation, and test sets, consisting of 436, 181, and 444 images, respectively.

However, the dataset also has the following problems: the docking ships are densely distributed, and the coincidence of the label frames is high, making it difficult to separate a single vessel; The background of the RS image is complex, and the similarity between the texture of the ship to be measured and the nearshore is large. The number of ships of each class is small, and the small sample size leads to insufficient learning and training, and poor robustness; Some images have cloud occlusion issues, and data quality is uneven. Therefore, we chose ships with relatively difficult data acquisition and a low number as the research object.

Furthermore, to enhance sample diversity, prevent overfitting during the training process, and improve robustness, data augmentation is applied to the original data, including operations such as image flipping, image rotation, adding noise, changing brightness, altering contrast, modifying saturation, and performing histogram equalization, among others. The choice of augmentation techniques is made based on specific circumstances.

3.2 Evaluation Indexes

The IOU is used to evaluate the degree of overlap between two bounding boxes [17]. It requires a ground truth bounding box (Bgt) and predicted bounding box (Bp) [18]. As shown in Eq. 1, by applying the IOUs, we can tell whether the prediction is true positive or false positive.

$$IOU = \frac{\text{area}(B_p \cap B_{gt})}{\text{area}(B_p \cup B_{gt})} \tag{1}$$

Commonly used quantitative evaluation indicators for object detection include precision P, recall R [19], mean average accuracy mAP, etc. As shown in Eq. 2, precision refers to the proportion of correctly classified samples in all samples, reflecting the model's ability to identify relevant targets. As shown in Eq. 3, recall reflects the model's ability to find the true regression box (i.e., the box labeled by the label).

$$Precision = \frac{TP}{TP + FP} \tag{2}$$

$$Recall = \frac{TP}{TP + FN} \tag{3}$$

where TP refers to the number of positive cases correctly classified by the classifier as positive cases when the iou is greater than or equal to the threshold, FP refers to the number of positive cases that the classifier misclassifies - predicts as positive cases, and FN refers to the number of positive cases that the classifier mistakenly classifies positive cases as negative cases when the iou is less than the threshold.

AP (Average Percision) is the average accuracy, R is the horizontal axis, P is the vertical axis to draw the PR curve, AP is the area under the PR curve of a specific type of all pictures, which can measure the quality of a category of target detection [20]. mAP refers to the average of all categories of APs in all images, and the higher the map, the higher the prediction accuracy of the model. mAP@0.5 refers to the mAP when the iou threshold is taken as 0.5, and mAP@0.5:0.95 refers to the average of the model under each iou threshold.

3.3 Experimental Analysis

As shown in Fig. 2, a total of 10,000 pseudo-ship RS images were generated. It is obvious that our generated vessel samples are well shaped and unified style, which can better satisfy the requirement of model training. In addition, we also put the generated pseudo-ship images into the target detection model trained with real ship RS images, and obtained high accuracy and recall rate, indicating that the pseudo-ship images generated by us are highly similar to the real ship images. The test results are shown in Table 1.

Table 1. Verification accuracy of the sample image.

Image	P	R	mAP50	mAP50-95
1	1.000	0.769	0.865	0.828
2	0.990	1.000	0.995	0.977
3	1.000	0.788	0.879	0.788
4	0.997	1.000	0.995	0.897
5	1.000	0.924	0.995	0.826
6	0.833	0.833	0.903	0.763
7	1.000	0.590	0.749	0.528
8	0.833	1.000	0.995	0.857
9	1.000	1.000	0.995	0.924
10	0.778	0.875	0.892	0.758

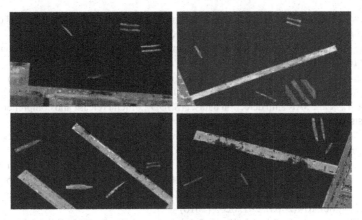

Fig. 2. The generated pseudo-ship images.

4 Contribution and Future Work

In this paper, we propose a novel ship data augmentation method that can augment ship samples with a small number of samples. Additionally, based on this method, a high-resolution remote sensing image dataset for fine-grained ship detection (HRSC-warcraft10k) was created, which comprising 10,000 images. We employ the mask to separate the ship from the background, and Style GAN was used to learn the characteristics of the ship and expand the ship data. Finally, the generated data was randomly fused with the background to obtain a large number of pseudo-RS images. Experiments show that the ship generated by our method is closed to the real acquired ship target. Besides, It can be well identified by the ship target.

Acknowledgments. This work is supported by National Key R&D Program of China (No.·2022YFB3902300) and the Fundamental Research Funds for the Central Universities, China (Grant No. 2042022dx0001).

Disclosure of Interests. The authors have no competing interests.

References

1. Ball, J.E., Anderson, D.T., Chan, C.S.: Comprehensive survey of deep learning in remote sensing: theories, tools, and challenges for the community. J. Appl. Remote Sens. **11**(4), 042609–042609 (2017)
2. Ma, L., et al.: Deep learning in remote sensing applications: a meta-analysis and review. ISPRS J. Photogrammetry Remote Sens. **152**, 166–177 (2019)
3. Signoroni, A., et al.: Deep learning meets hyperspectral image analysis: a multidisciplinary review. J. Imaging **5**(5), 52 (2019)
4. Kanjir, U., Greidanus, H., Oštir, K.: Vessel detection and classification from spaceborne optical images: a literature survey. Remote Sens. Environ. **207**, 1–26 (2018)
5. Arnesen, T.N., Olsen, R.: Literature review on vessel detection (2004)

6. Al-Rawi, M., Qutaishat, M., Arrar, M.: An improved matched filter for blood vessel detection of digital retinal images. Comput. Biol. Med. **37**(2), 262–267 (2007)
7. Sheng, K., et al.: Research on ship classification based on trajectory features. J. Navig. **71**(1), 100–116 (2018)
8. Huang, L., et al.: Multiple features learning for ship classification in optical imagery. Multimedia Tools Appl. **77**, 13363–13389 (2018)
9. Zhang, T., et al.: SAR ship detection dataset (SSDD): official release and comprehensive data analysis. Remote Sens. **13**(18), 3690 (2021)
10. Wang, Y., et al.: A SAR dataset of ship detection for deep learning under complex backgrounds. Remote Sens. **11**(7) (2019)
11. Wah, C., Branson, S., Welinder, P., Perona, P., Belongie, S.: The Caltech-UCSD birds-200-2011 dataset. Technical report CNS-TR-2011-001, California Institute of Technology (2011)
12. Mallat, S.: Understanding deep convolutional networks. Philos. Trans. Royal Soc. A Math. Phys. Eng. Sci. **374**(2065), 20150203 (2016)
13. Creswell, A., et al.: Generative adversarial networks: an overview. IEEE Sig. Process. Mag. **35**(1), 53–65 (2018)
14. Liu, Z., Wang, H., Weng, L., Yang, Y.: Ship rotated bounding box space for ship extraction from high-resolution optical satellite images with complex backgrounds. IEEE Geosci. Remote Sens. Lett. **13**, 1074–1078 (2016)
15. Karras, T., Laine, S., Aila, T.: A style-based generator architecture for generative adversarial networks. In: Proceedings of the IEEE/CVF Conference on Computer Vision and Pattern Recognition (2019)
16. Porter, T., Duff, T.: Compositing digital images. In: Proceedings of the 11th Annual Conference on Computer Graphics and Interactive Techniques (1984)
17. Everingham, M., et al.: The pascal visual object classes (voc) challenge. Int. J. Comput. Vis. **88**, 303–338 (2010)
18. Papageorgiou, C., Poggio, T.: A trainable system for object detection. Int. J. Comput. Vis. **38**, 15–33 (2000)
19. Davis, J., Goodrich, M.: The relationship between Precision-Recall and ROC curves. In: Proceedings of the 23rd International Conference on Machine Learning (2006)
20. Yang, G., Chen, B.: Application of improved YOLOv5 object detection model in urban street scenes. J. Comput. Appl. (2023)

Applications of Deep Learning in Satellite Communication: A Survey

Yuanzhi He[1,2]([✉]), Biao Sheng[1], Yuan Li[2], Changxu Wang[2], Xiang Chen[1], and Jinchao Liu[3]

[1] School of Systems Science and Engineering, Sun Yat-Sen University, Guangzhou 100876, China
he_yuanzhi@126.com
[2] Institute of Systems Engineering, Academy of Military Sciences, Beijing 100141, China
[3] China Coast Guard, Beijing 100141, China

Abstract. Satellite communication is a key aspect of future 6G networks, and the impact of artificial intelligence technology utilizing deep learning on satellite communications has garnered significant interest. This paper outlines the current research status of deep learning applications in satellite communication from the perspective of the physical layer, data link layer, and network layer. It also examines the limitations of deep learning in satellite communication applications and anticipates potential research directions for the future.

Keywords: Satellite communication · Deep learning · Physical layer · Data link layer · Network layer

1 Introduction

In recent years, significant advancements have been made in the construction of satellite communication (SatCom) systems, leading to the emergence of new concepts, technologies, and projects. The low-earth orbit (LEO) constellations represented by Starlink [1] and Hongyan [2] are under construction, which promote the development of a series of satellite communication networks with different focuses, such as Internet of satellite (IoS) [3], space-ground integration network (SGIN) [4,5], integrated satellite-terrestrial networks (ISTN/ISN) [6], and direct cell-to-satellite communication [7]. Figure 1 shows a typical space-air-ground-sea integrated satellite communication network, including satellites, aircraft, unmanned aerial vehicles, pseudo satellites, vehicles, ships, etc.

In June 2023, the International Telecommunication Union (ITU) approved the IMT-2030 (6G) vision framework [8]. Among them, AI-related capabilities and full coverage are the important communication capabilities proposed in IMT-2030, as shown in Fig. 2. Satellite communication is an important way to connect remote areas, the ocean, and the air. The integration of artificial intelligence and satellite communication is inevitable under the vision of ITU. These emerging

Q. Yu (Ed.): SINC 2023, CCIS 2057, pp. 17–33, 2024.
https://doi.org/10.1007/978-981-97-1568-8_3

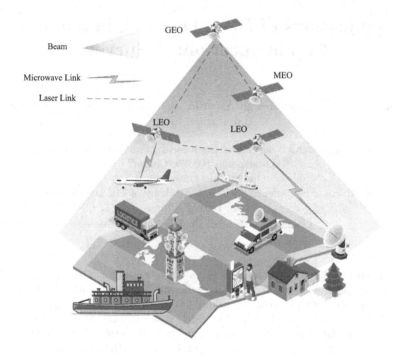

Fig. 1. Architecture of space-air-ground-sea integrated information network.

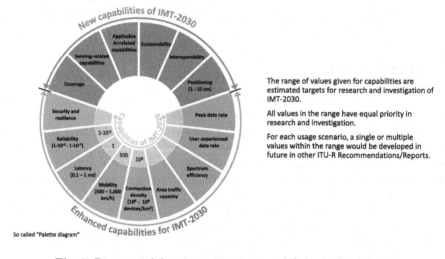

Fig. 2. Diagram of the communication capabilities in IMT-2030.

technologies have been widely concerned by researchers, and the role and status of satellite communication are becoming more and more important.

With the construction of new satellite communication networks such as SGIN, many new problems and difficulties related to satellite communication have also emerged. For example, wireless resource allocation and routing become more difficult in the SGIN than that in a single satellite network. How to solve the various problems better is the key to improve the quality of satellite communication service.

Deep learning (DL) is a fundamental technology in artificial intelligence, capable of efficiently training a deep neural network (DNN) with extensive data. The DNN can extract data features and tackle different classification and decision optimization tasks. Deep learning is used in image classification, speech recognition, games like Go, and various other fields. Its effectiveness in solving classification and sequential decision problems provides opportunities for addressing similar issues in satellite communication.

In this paper, the applications of deep learning in satellite communications will be studied in different communication levels, and the main contributions of this paper includes:

Both satellite communication and deep learning are developing rapidly, and this paper tracks the latest development of deep learning in satellite communications.

This paper analyzes the applications of deep learning in satellite communications from the perspective of physical layer, link layer and network layer, which is helpful to understand the role of deep learning at different levels. Most of the existing reviews on satellite communication are aimed at a specific application or method. For example, reviews on access [9], or reviews on the application of deep reinforcement learning in satellite communication, or the focus is not on the relationship between deep learning and satellite communication [10], and these literatures rarely consider the relationship between deep learning and satellite communication from a hierarchical perspective [11].

This paper focuses on the network model, hardware, and training data that are very important in deep learning. There's very little literature focus on this. The neural network applied to satellite communication is usually a classical network structure, the training data is mainly simulation data, and the hardware is not given in detail.

To better understand the applications of deep learning in satellite communication, this paper first classifies them from the perspective of physical layer, data link layer, and network layer, and expounds on the category of deep learning technology used in different communication layers, as shown in Fig. 3. Then, the status of satellite network types, deep learning methods, neural network models, datasets, and hardware information in the research are summarized. Finally, some problems and possible development directions of deep learning in enabling satellite communication are pointed out.

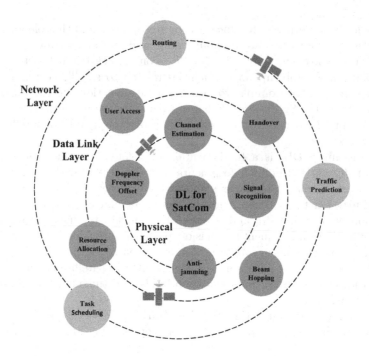

Fig. 3. Typical applications of deep learning in satellite communication.

2   Physical Layer

The role of the physical layer is to enable the signal to obtain reliable transmission on the physical channel. Signal processing at the transmitter includes source coding, channel coding, modulation, etc., and signal recovery, channel estimation, and other processing at the receiver. Deep learning has been applied to channel estimation [12], signal identification [13], interference suppression [14], Doppler frequency offset [15] and other problems at the physical layer.

2.1   Channel Estimation

In satellite communication, information is transmitted in space by electromagnetic waves. The space environment is complex and changeable, and the electromagnetic wave will produce path loss, rain attenuation, fog attenuation, and other damages in the transmission process, which leads to the difference between the transmitted signal and the received signal. Therefore, channel estimation for satellite communication has important value.

In LEO satellite communication systems, obtaining accurate channel state information (CSI) is crucial to achieve high performance. Least squares (LS) channel estimation is a simple traditional channel estimation scheme, but it does not consider the compensation of channel estimation error. Kang et al. [16] proposed a denoising convolutional neural network (DnCNN) based on deep learning for channel estimation of massive multiple-input single-output LEO satellite communication systems. Numerical results show that the DnCNN method can effectively improve the accuracy of the LS channel estimator in estimating the channel.

Zhang et al. [17] proposed a CSI prediction scheme based on deep learning to solve the channel aging problem of massive multiple-input multiple-output (mMIMO) LEO satellite communication system by exploiting the correlation of varying channels. In the paper, a satellite channel predictor composed of long short-term memory (LSTM) units is designed. The predictor is first trained by offline learning, and then the corresponding output results are fed back online according to the input data to realize the channel feature extraction and future CSI prediction in LEO satellite scenarios. Numerical results show that the proposed deep learning-based predictor can effectively alleviate the channel aging problem in LEO satellite mMIMO systems.

Aiming at the issue of high altitude platform station LEO (HAPS-LEO) satellite communication network channel being impacted by time-varying conditions, Guven et al. [18] introduced a deep learning method to address the channel estimation challenge. The paper presents a sequential channel and carrier frequency offset estimation using a convolutional neural network (CNN), demonstrating the CNN's effectiveness in minimizing estimation errors and enhancing mean square error and bit error rate performances.

2.2   Signal Recognition

Compared with terrestrial communication, satellite communication has lower received power and poorer signal quality. Enhanced signal detection and identification technology can improve the quality of satellite communication, making it a crucial aspect of satellite communication.

Zha et al. [19] introduced a signal recognition and demodulation model leveraging a recurrent neural network (RNN), which is LSTM. The LSTM was utilized to directly extract the underlying features of signal timing, while a fully connected neural network (FCNN) was employed to map these features' dimensions, and finally the modulation recognition and demodulation of the target signal were completed. The method does not need to estimate the carrier to noise ratio of the target signal, overcomes the defect of artificially determining the threshold, and has a strong tolerance for signal frequency offset error and timing error. Simulation results show that when the network training reaches the steady state, the target signal recognition rate is close to 98%, and the demodulation error rate is close to the theoretical threshold under the condition of signal-to-noise ratio (SNR) of 6 dB.

Ren et al. [20] proposed an RNN-based algorithm for satellite modulation signal recognition, which takes the IQ data of the signal as the input of the model, extracts time-sharing features through LSTM, classifies the fully connected layer, and finally completes the recognition. When the sampling length is 512 and SNR is greater than 4dB, the recognition rate approaches 100%.

Compared with the K-nearest neighbor classifier, the LSTM network has better recognition performance, especially in the case of low SNR, it can recognize multiple modulation modes efficiently activity.

2.3   Anti-jamming

In satellite communication, the received signal power is relatively low, resulting in a low signal-to-noise ratio. Therefore, the anti-jamming ability is crucial for satellite communication.

For the anti-jamming issue in the heterogeneous IoS, Han et al. [21] researched a spatial anti-jamming approach to minimize routing costs through a Stackelberg game and deep reinforcement learning. The problem is structured as a hierarchical anti-jamming Stackelberg game, with two main phases: route selection and quick anti-jamming decision. A deep reinforcement learning routing algorithm (DRLR) is used to obtain the available route subset, while a fast response anti-jamming algorithm (FRA) is employed for rapid decision-making. By analyzing jammer strategies with DRLR and FRA algorithms, users can adaptively make anti-jamming decisions according to the jamming environment. Simulation results demonstrate the algorithm's effectiveness in reducing routing costs and enhancing anti-interference performance compared with current methods.

2.4   Doppler Frequency Offset

The Doppler frequency offset induced by the highly dynamic characteristics of LEO satellites complicates signal recovery at the receiver.

To address this issue in LEO satellite communication systems, Li et al. [22] proposed a Doppler frequency offset pre-compensation algorithm using multimodal long short-term memory (MLSTM-DPC). This algorithm selects single or multiple LSTM models based on the discrepancy between the current ephemeris and time to predict orbit parameters. These predicted parameters are then utilized to infer orbits, ultimately predicting the Doppler frequency offset pre-compensation value. Simulation results demonstrate a 36.39% enhancement in the effectiveness of the MLSTM-DPC algorithm in mitigating frequency offset, along with a significant reduction in calculation time.

2.5   Summary

Deep learning algorithms help to acquire channel knowledge and predict time-varying channels at the physical layer. The algorithm mainly focuses on the relationship between input and output in wireless channels and generally uses supervised learning(SL) methods. Deep learning algorithms perform better in classification problems and are more robust to unpredictable errors because they can learn it if the dataset is complete. It has more advantages than traditional methods in signal recognition.

Deep learning demands significant computing power, advanced algorithms, and extensive data. Table 1 provides a concise overview of literature focusing on satellite network types, deep learning methods, neural network models, datasets, and hardware information. The analysis reveals a diverse range of satellite network types. The predominant deep learning approach is supervised learning, with neural network models primarily utilizing classical CNN and LSTM architectures. However, the datasets predominantly consist of simulated data, which relatively lacks authority, reusability, and standardization. Additionally, many papers do not specify the CPU or GPU used for training.

Table 1. Typical applications of deep learning in the physical layer.

Applications	Satellite Networks	Deep Learning Methods	Neural Networks	Open Source Datasets	Hardware Information	Ref.
Channel Estimation	LEO	SL	DnCNN	No	/	[16]
	LEO	SL	LSTM	No	/	[17]
	HAPS-LEO	SL	CNN	No	/	[18]
Signal Recognition	DVB-S2	SL	LSTM	No	/	[19]
	Non-specific	SL	LSTM	No	/	[20]
Anti-Jamming	IoS	DRL	LSTM	No	/	[21]
Doppler Frequency Offset	LEO	SL	LSTM	No	CPU	[22]

3   Data Link Layer

The role of the data link layer is to provide access, error detection, multiplexed data streams for user data, and provide reliable data connection services for users. Deep learning has been applied in resource allocation [23], user access [24], handover [25], beam hopping [26], and other problems.

3.1   Resource Allocation

Limited by its platform, satellite communication can use limited bandwidth and power resources, so the resource allocation algorithm is very important to the quality of satellite communication. Deep learning, especially deep reinforcement learning, is relatively widely used in the field of resource allocation.

Due to the limited resources of the space-ground integrated satellite network, how to effectively allocate the resources has become a big challenge. Li et al. [27] proposed a non-orthogonal multiple access (NOMA) based resource allocation framework for terrestrial satellite networks. In the proposed framework, multi-agent deep deterministic policy gradient (MADDPG) method was used to achieve maximum energy efficiency through user association, power control, and cache design. Simulation results show that the proposed method has better optimization performance than the traditional single-agent deep reinforcement learning algorithm, and can effectively solve the problem of resource allocation and cache design in the space-ground integrated network.

With the expansion of multibeam satellite (MBS) network scale, how to efficiently and dynamically allocate scarce bandwidth and spectrum resources while ensuring user quality of service (QoS) has become a big challenge. Ma et al. [28] designed a dynamic bandwidth allocation framework using proximal policy optimization (DBA-PPO) to meet time-varying traffic demand, maximize utilization, and guarantee QoS of users in the MBS system. Experimental results show that the proposed bandwidth allocation algorithm can flexibly achieve the desired effect with lower complexity and is more cost-effective for large-scale MBS communication scenarios.

3.2 User Access

In the field of satellite access, at present, NOMA [29,30] related schemes have attracted more attention from researchers and are considered possible access schemes for 6G. However, the access enabled by deep learning has also been studied accordingly.

Zhang et al. [31] examined the user pairing issue in the power domain non-orthogonal multiple access scheme in satellite networks. It is assumed that various satellite applications possess diverse delay QoS needs, and the notion of effective capacity is utilized to describe the influence of delay constraints on the attained performance. The objective is to choose users for creating NOMA user pairs and optimize resource utilization. To this end, the power allocation factor is first obtained by ensuring that the capacity achieved by the delay-sensitive users is not less than that achieved by the orthogonal multiple access (OMA) scheme. Then, considering the non-convex user selection in the delay-constrained NOMA satellite network, the DRL algorithm was used for dynamic user selection. The channel conditions and delay requirements of the users are considered as states, and the DRL algorithm is used to search for the best user that can achieve the maximum performance under the power allocation factor, which is paired with the delay-sensitive user to form a NOMA user pair. Simulation results show that the proposed DRL-based user selection scheme can output the optimal action in each time slot, thus providing superior access performance than the random selection strategy and OMA scheme.

LEO satellite networks exhibit extremely long link distances for many users under time-varying network topologies. This makes existing multiple access protocols unsuitable. To overcome this problem, Lee et al. [32] proposed a contention-based random access solution called emergency random access channel protocol (eRACH). It emerges by interacting with a non-stationary network environment using multi-agent deep reinforcement learning. By exploiting known satellite orbit patterns, eRACH does not require central coordination or additional communication between users. Simulation results show that the average network throughput of the proposed eRACH is 54.6% higher, and the average access delay is reduced by about two times.

3.3   Handover

Since the large-scale construction of LEO constellation communication by Starlink, more and more LEO satellite communication systems have been proposed and are under steady construction. In the process of satellite communication, a single communication link is difficult to maintain for a long time, and users need to switch in different beams.

Yang et al. [33] proposed a handover method of DQN framework with momentum adaptive learning rate (DQN-ALRM), which can not only improve the decision-making accuracy, but also improve the learning efficiency. The customized DQN framework can solve the problem of large-size state space, and the proposed DQN-ALRM method can adjust the learning rate at any time according to the training error situation. Simulation results show that the proposed method has advantages in convergence speed, handover rate, call failure rate, and multi-index QoS.

Wang et al. [34] proposed a new handover scheme based on DRL by simultaneously considering multiple handover factors such as handover signaling overhead, remaining visible time, received signal strength, shortest distance, and satellite load balance. Simulation results show that the proposed DRL-based handover scheme can reduce the number of handovers by more than 21% compared with the comparison baseline in the case of no handover failure.

Leng et al. [35] introduced a multi-attribute handover strategy considering satellite buffer capacity, remaining service time, and idle channel availability. They also presented a cache-aware intelligent handover strategy using DRL to optimize the system's long-term benefits. Compared with the traditional strategy, the proposed strategy can reduce the handover failure rate up to 81% when the system buffer occupancy rate reaches 90%, and has a lower call blocking probability in the multi-user arrival scenario.

Xu et al. [36] proposed a user-centric intelligent handover mechanism for mobile satellite networks, which selects the access satellite by predicting the service time and communication channel resources. Deep reinforcement learning is used to maximize the quality of experience of the user terminal through the predicted handover factors. Simulation results show that the proposed handover mechanism has good performance in terms of handover time, success rate, and end-to-end delay.

3.4   Beam Hopping

Beam-hopping is a flexible beam scheduling method envisaged by multibeam satellites to improve system throughput.

Lei et al. [37] developed a deep learning-assisted approach to facilitate efficient beam hopping in multibeam satellite systems. Adopting beam hopping can provide a high degree of flexibility to manage irregular and time-varying traffic requests within the satellite coverage area. This paper proposes a method combining learning and optimization to provide fast, feasible, and near-optimal solutions for beam hopping scheduling. Numerical studies show that the learning component can greatly accelerate the process of beam hopping mode selection and allocation, while the optimization component can guarantee the feasibility of the solution and improve the overall performance.

3.5 Summary

Because deep reinforcement learning has a good effect on serialization decisions, it has been applied to many satellite communication tasks that require decision optimization, such as resource allocation, user access, handover, and beam hopping. Deep reinforcement learning technology can learn the relationship between input and output on a large amount of data, and use the reward to find the best optimization scheme.

Similar to Table 1, Table 2 displays statistics on satellite network types, deep learning methods, neural network models, datasets, and hardware information. It is clear that there is a range of satellite network types, and the methods are mainly focused on deep reinforcement learning. The datasets utilized are all acquired through simulation, and the CPU or GPU employed for training is not specified in the literature. In comparison to Table 1, the neural network model may use a fully connected layer, and the dataset also presents challenges such as lack of standardization, authority, and reusability.

Table 2. Application status of deep learning in data link layer.

Applications	Satellite Networks	Deep Learning Methods	Neural Networks	Open Source Datasets	Hardware Information	Ref.
Resource Allocation	Terrestrial-Satellite networks	MADDPG	/	No	/	[27]
	Multi-Beam Satellite	DBA-PPO	FCNN	No	/	[28]
User Access	NOMA-Based Satellite Networks	DRL	/	No	/	[31]
	LEO	MADRL	FCNN	No	/	[32]
Handover	SGIN	DRL	/	No	/	[33]
	LEO	DRL	Conv2D	No	/	[34]
	LEO	DRL,DQN	/	No	/	[35]
	Mobile Satellite Networks	DRL	/	No	/	[36]
Beam Hopping	LEO	DL	FC-DNN	No	/	[37]

4   Network Layer

The network layer has the role of connecting different networks, deciding the best route, and managing network traffic. The role of deep learning in the network layer includes routing optimization [38], traffic prediction [39], task scheduling [40], etc.

4.1   Routing

Routing plays a crucial role in the network layer. A flawed routing algorithm can result in network congestion and various other issues that need to be handled with care.

Liu et al. [41] conceive space-air-ground integrated networks(SAGIN) supporting maritime communications, in which LEO satellite constellations, passenger aircraft, ground base stations, and ships serve as space, air, ground, and sea layers, respectively. In order to meet the heterogeneous service requirements and adapt to the time-varying and self-organizing nature of SAGIN, a deep learning assisted multi-objective routing algorithm was proposed, which utilized the quasi-predictive network topology and operated in a distributed manner. Simulation results based on real satellite, flight, and shipping data in the North Atlantic region show that the proposed deep learning-assisted multi-objective routing algorithm can achieve near-Pareto optimal performance.

Wang et al. [42] combined software-defined network, AI technology, and fuzzy logic to optimize the multi-task routing strategy in the ISN. The geostationary earth orbit (GEO) controller collects the load information of ISTN at different moments, and the ground computing center collects historical traffic data from the GEO controller for CNN model training and updating. The GEO satellite utilizes the trained CNN model to make routing decisions. Considering that the judgment of CNN may contradict the user requirements, fuzzy logic is used to evaluate the task requirements to improve the output of CNN to obtain the best routing policy. Simulation results show that the multi-task routing method based on fuzzy CNN has better performance under different conditions.

4.2   Traffic Prediction

Traffic prediction is important in many satellite applications such as congestion control, dynamic routing, dynamic channel assignment, network planning, and network security.

Wan et al. [43] proposed a traffic classification method based on deep packet inspection and CNN and verified it with open datasets. Experimental results show the effectiveness of the proposed method.

Zhu et al. [44] proposed a LSTM model with attention mechanism for traffic prediction. Considering that the input and output of traffic prediction are a sequence, the proposed model can balance the influence of different parts of the input on the output. Simulation results demonstrate a significant enhancement in prediction accuracy compared with the auto-regressive integrated moving average model, random forest, and traditional RNN.

4.3 Task Scheduling

Task scheduling is a common task in the network layer, especially the scheduling of computing tasks is the focus of research.

Zhang et al. [45] examined the scheduling problem of computing tasks in SAGIN for remote Internet of Things (IoT). The aim is to design an offloading strategy for each unmanned aerospace vehicle (UAV) to maximizes task quantity and reduce energy consumption. To adapt the complex and changing environment, they proposed a task offloading method based on the Actor-Critic framework. The actor observes the environment and outputs the current offloading decision, and the critic evaluates the behavior of the actor and coordinates the behavior of all UAVs to increase the reward of the system. Simulation results show that the proposed offloading strategy has fast convergence speed, increases the number of tasks, and improves the energy utilization of the UAV.

Lan et al. [46] proposed an ISTN to achieve satellite-assisted mission unloading under dynamic conditions. A privacy protection algorithm based on deep reinforcement learning is introduced to achieve the optimal unloading strategy. Experimental results show that the proposed algorithm is superior to other benchmark algorithms in completion time, energy consumption, privacy protection and communication reliability.

Zhang et al. [47] studied the problem of computing task offloading and resource allocation in multi-layer LEO satellite networks assisted by UAVs. In order to minimize the weighted sum of energy consumption and delay in the system, the problem was formulated as a constrained optimization problem, and then it was transformed into a Markov decision problem, and a task offloading and resource allocation algorithm based on deep deterministic policy gradient and long short-term memory was proposed. Simulation results show that the proposed solution outperforms baseline methods, and the proposed framework and algorithm have the potential to provide reliable communication services in emergencies.

Han et al. [48] proposed an ISTN architecture to support delay-sensitive task offloading in remote IoT. In order to minimize the overall task offloading delay, a hierarchical Markov decision process framework was established, and an algorithm based on hybrid proximal policy optimization (H-PPO) was further developed. The proposed algorithm designs a hybrid actor-critic architecture to deal with mixed discrete and continuous actions. Simulation results verify the superiority of the proposed ISTN architecture and the H-PPO-based algorithm.

4.4 Summary

The network layer primarily utilizes deep reinforcement learning for routing optimization and task offloading, while supervised learning is employed for traffic prediction tasks.

In Table 3, it is evident that satellite networks in the network layer encompass a variety of types, with a notable emphasis on the space-ground integrated network. This network type, characterized by a high node count, intricate topology, and substantial user traffic, presents significant challenges for routing optimization. Deep Learning techniques predominantly revolve around supervised learning and deep reinforcement learning, leveraging various neural network models like CNN, LSTM, and FCNN. These models are comparatively less intricate than those used in terrestrial communication or computer science applications. Some researchers utilize adapted open-source datasets for experimentation in satellite communications. Most papers do not disclose the CPU or GPU used for training.

Table 3. Application status of deep learning in network layer.

Applications	Satellite Networks	Deep Learning Methods	Neural Networks	Open Source Datasets	Hardware Information	Ref.
Routing	SAGINs	DL	DNN	Yes	/	[41]
	ISN	SL	CNN	Yes	/	[42]
Traffic Prediction	Non-specific	SL	CNN	Yes	CPU	[43]
	LEO	SL	LSTM+ Attention	Yes	/	[44]
Task Scheduling	SAGIN for remote IoT	DRL	FCNN	No	GPU	[45]
	ISTN	DRL	/	No	GPU	[46]
	UAV- Assisted LEO Satellite	DRL	LSTM	No	/	[47]
	ISTN for IoT	DRL	FCNN	No	/	[48]

5 Conclusion

Deep learning and satellite communication technology are developing rapidly, and satellite communication technology powered by deep learning brings a better experience for space-ground integrated networks. The application of deep reinforcement learning in resource allocation, handover, and routing control has been widely studied. Meanwhile, supervised learning plays an important role in signal recognition, and traffic prediction. However, deep learning still encounters challenges in satellite communication:

1. Dataset problem. Using simulation data to verify the effect of communication system is the mainstream solution, perhaps because the network structure is always different, and makes it difficult to establish a general dataset. There is no standard open-source dataset related to satellite communication, and the lack of datasets restricts the research progress of deep learning for satellite

communication. It is of great significance to increase the construction of open datasets for satellite communication.

2. Neural network fragmentation problem. When solving different problems, different types of deep learning algorithms and neural network models need to be used. This results in the creation of multiple neural network fragments, leading to substantial storage space consumption.
3. Computing power problem. Deep learning demands significant computing power, with larger models requiring even more power. Handheld devices or satellites often struggle to provide such power. Currently, many research efforts overlook the influence of neural network size in satellite communication applications, especially the competition between satellite computing power and communication power consumption.
4. Practical application problem. Compared with the practice of deep learning in Go, automatic driving, and UAV flight, there is no practice of deep learning in satellite communication at present, and it is often verified by simulation. Deep learning-based satellite communication needs more practical implementation.

Solving the aforementioned problems can effectively promote the application research of deep learning in satellite communication. In addition, strengthening the integration of deep learning and satellite communication at the hardware level, for example, carrying out the end-to-end physical layer design of satellite communication [49,50] from the perspective of system theory, could foster the growth of intelligent internal networks. At the same time, based on the concept of AI for science, it is very valuable to conduct basic theory research based on deep learning in satellite communication, rather than limiting it to the engineering application level.

References

1. Duan, T., Dinavahi, V.: Starlink space network-enhanced cyber-physical power system. IEEE Trans. Smart Grid **12**(4), 3673–3675 (2021)
2. Wang, Z.J., Du, X.J., Yin, J.W., et al.: Development and prospect of LEO satellite Internet. Appl. Electron. Tech. **46**(7), 49–52 (2020)
3. Wang, P., Zhu, S., Li, C., et al.: Analysis on development of satellite internet standardization. Radio Commun. Technol. **49**(5), 1–7 (2023)
4. Wang, C.T., Zhai, L.J., Xu, X.F.: Development and prospects of space-terrestrial integrated information network. Radio Commun. Technol. **46**(5), 493–504 (2020)
5. Fang, X., Feng, W., Wei, T., et al.: 5G embraces satellites for 6G ubiquitous IoT: basic models for integrated satellite terrestrial networks. IEEE Internet Things J. **8**(18), 14399–14417 (2021)
6. Zhang, S.J., Zhao, X.T., Zhao, Y.F., et al.: Integration of satellite internet and terrestrial networks: integrated mode, frequency usage and application prospects. Radio Commun. Technol. **49**(5) (2023)
7. Sun, Y.H., Peng, M.G.: Low earth orbit satellite communication supporting direct connection with mobile phones: key technologies, recent progress and future directions. Telecommun. Sci. **39**(2), 25–36 (2023)

8. ITU-R, Workshop on IMT for 2030 and beyond. https://www.ituint/en/ITU-R/study-groups/rsg5/rwp5d/imt-2030/Pages/default.aspx
9. Xiao, Z., et al.: LEO satellite access network (LEO-SAN) towards 6G: challenges and approaches. IEEE Wirel. Commun. 1–8 (2022)
10. Azari, M.M., Solanki, S., Chatzinotas, S., et al.: Evolution of non-terrestrial networks from 5G to 6G: a survey. IEEE Commun. Surv. Tutor. **24**(4), 2633–2672 (2022)
11. Fourati, F., Alouini, M.S.: Artificial intelligence for satellite communication: a review. Intell. Converg. Netw. **2**(3), 213–243 (2021)
12. Wang, X., Shen, W., Xing, C., et al.: Joint Bayesian channel estimation and data detection for OTFS systems in LEO satellite communications. IEEE Trans. Commun. **70**(7), 4386–4399 (2020)
13. Yan, W.K., Yan, Y., Fan, Y.N., Yao, X.J., Gao, X., Sun, W.: A modulation recognition algorithm based on wavelet transform entropy and high-order cumulant for satellite signal modulation. Chin. J. Space Sci. **241**(6), 968–975 (2021). (in Chinese)
14. Li, J., Tang, X., Gao, L., Chen, L.: Satellite communication anti-jamming based on artificial bee colony blind source separation. In: 2021 6th International Conference on Communication, Image and Signal Processing (CCISP), pp. 240–244 (2021)
15. Subramanian, V., Karunamurthy, J.V., Ramachandran, B.: Hardware doppler shift emulation and compensation for LoRa LEO satellite communication. In: 2023 International Conference on IT Innovation and Knowledge Discovery (ITIKD), pp. 1–6 (2023)
16. Kang, M.J., Lee, J.H., Chae, S.H.: Channel estimation with DnCNN in massive MISO LEO satellite systems. In: 2023 Fourteenth International Conference on Ubiquitous and Future Networks (ICUFN), pp. 825–827 (2023)
17. Zhang, Y., Wu, Y., Liu, A., et al.: Deep learning-based channel prediction for LEO satellite massive MIMO communication system. IEEE Wirel. Commun. Lett. **10**(8), 1835–1839 (2021)
18. Güven, E., Kurt, G.K.: CNN-aided channel and carrier frequency offset estimation for HAPS-LEO links. In: 2022 IEEE Symposium on Computers and Communications (ISCC), pp. 1–6 (2022)
19. Zha, X., Peng, H., Qin, X., Li, T.Y., Li, G.: Satellite amplitude-phase signals modulation identification and demodulation algorithm based on the cyclic neural network. Acta Electron. Sin. **11**(47), 2443–2448 (2019)
20. Ren, J., Ji, L.B., Dang, L.: Satellite signal modulation recognition algorithm based on deep learning. Radio Eng. **52**(4), 529–535 (2022)
21. Han, C., Huo, L., Tong, X., et al.: Spatial anti-jamming scheme for internet of satellites based on the deep reinforcement learning and stackelberg game. IEEE Trans. Veh. Technol. **69**(5), 5331–5342 (2020)
22. Li, H., Liu, Y., Shi, J., Zhou, Y., Zhuo, R., Li, S.: Multimodal LSTM forecasting for LEO satellite communication terminal access. In: 2023 IEEE 97th Vehicular Technology Conference (VTC2023-Spring), pp. 1–5 (2023)
23. Liao, X., Hu, X., Liu, Z., et al.: Distributed intelligence: a verification for multi-agent DRL based multibeam satellite resource allocation. IEEE Commun. Lett. **42**(12), 2785–2789 (2020)
24. Wu, X.W., Ling, X., Zhu, L.D.: Access and mobility management technologies for 6G satellite communications network. Telecommun. Sci. **37**(6), 78–90 (2021)
25. Jiang, Z., Li, W., Wang, X., et al.: A LEO satellite handover strategy based on graph and multiobjective multiagent path finding. Int. J. Aerosp. Eng. **2023**, 1–16 (2023)

26. Hu, X., Zhang, Y., Liao, X., et al.: Dynamic beam hopping method based on multi-objective deep reinforcement learning for next generation satellite broadband systems. IEEE Trans. Broadcast. **66**(3), 630–646 (2020)
27. Li, X., Zhang, H., Zhou, H., et al.: Multi-agent DRL for resource allocation and cache design in terrestrial-satellite networks. IEEE Trans. Wireless Commun. **22**(8), 5031–5042 (2023)
28. Ma, S., Hu, X., Liao, X., Wang, W.: Deep reinforcement learning for dynamic bandwidth allocation in multi-beam satellite systems. In: 2021 IEEE 6th International Conference on Computer and Communication Systems (ICCCS), pp. 955–959 (2021)
29. Makki, B., Chitti, K., Behravan, A., et al.: A survey of NOMA: current status and open research challenges. IEEE Open J. Commun. Soc. **1**, 179–189 (2020)
30. Zhu, X., Jiang, C., Kuang, L., et al.: Non-orthogonal multiple access based integrated terrestrial-satellite networks. IEEE J. Sel. Areas Commun. **35**(10), 2253–2267 (2017)
31. Zhang, Q., An, K., Yan, X., et al.: User pairing for delay-limited NOMA-based satellite networks with deep reinforcement learning. Sensors **23**(16), 7062 (2023)
32. Lee, J.H., Seo, H., Park, J., et al.: Learning emergent random access protocol for LEO satellite networks. IEEE Trans. Wireless Commun. **22**(1), 257–269 (2023)
33. Yang, J., Xiao, Z., Cui, H., et al.: DQN-ALrM-based intelligent handover method for satellite-ground integrated network. IEEE Trans. Cogn. Commun. Netw. **9**(4), 977–990 (2023)
34. Wang, J., Mu, W., Liu, Y., Guo, L., Zhang, S., Gui, G.: Deep reinforcement learning-based satellite handover scheme for satellite communications. In: 2021 13th International Conference on Wireless Communications and Signal Processing (WCSP), pp. 1–6 (2021)
35. Leng, T., Xu, Y., Cui, G., et al.: Caching-aware intelligent handover strategy for LEO satellite networks. Remote Sens. **13**(11), 22–30 (2021)
36. Xu, H., Li, D., Liu, M., et al.: QoE-driven intelligent handover for user-centric mobile satellite networks. IEEE Trans. Veh. Technol. **69**(9), 10127–10139 (2020)
37. Lei, L., Lagunas, E., Yuan, Y., Kibria, M.G., Chatzinotas, S., Ottersten, B.: Deep learning for beam hopping in multibeam satellite systems. In: 2020 IEEE 91st Vehicular Technology Conference (VTC2020-Spring), pp. 1–5 (2020)
38. Cao, X., Li, Y., Xiong, X., et al.: Dynamic routings in satellite networks: an overview. Sensors **22**(12), 45–52 (2022)
39. Liu, Y., Yu, J.J.Q., Kang, J., et al.: Privacy-preserving traffic flow prediction: a federated learning approach. IEEE Internet Things J. **7**(8), 7751–7763 (2020)
40. Wu, G., Luo, Q., Zhu, Y., et al.: Flexible task scheduling in data relay satellite networks. IEEE Trans. Aerosp. Electron. Syst. **58**(2), 1055–1068 (2022)
41. Liu, D., Zhang, J., Cui, J., et al.: Deep learning aided routing for space-air-ground integrated networks relying on real satellite, flight, and shipping data. IEEE Wirel. Commun. **29**(2), 177–184 (2022)
42. Wang, F., Jiang, D., Wang, Z., et al.: Fuzzy-CNN based multi-task routing for integrated satellite-terrestrial networks. IEEE Trans. Veh. Technol. **71**(2), 1913–1926 (2023)
43. Wan, X., Fu, X., Li, J., et al.: Research on satellite traffic classification based on deep packet recognition and convolution neural network. In: 2023 8th International Conference on Computer and Communication Systems (ICCCS), pp. 494–498 (2023)

44. Zhu, F., Liu, L., Lin, T.: An LSTM-based traffic prediction algorithm with attention mechanism for satellite network. In: Proceedings of the 2020 3rd International Conference on Artificial Intelligence and Pattern Recognition, pp. 205–209 (2020)
45. Zhang, S., Liu, A., Han, C., et al.: Multi-agent reinforcement learning-based orbital edge offloading in SAGIN supporting internet of remote things. IEEE Internet Things J. **10**(23), 20472–20483 (2023)
46. Lan, W., Chen, K., Li, Y., et al.: Deep Reinforcement Learning for Privacy-Preserving Task Offloading in Integrated Satellite-Terrestrial Networks. arXiv (2023). http://arxiv.org/abs/2306.17183
47. Zhan, H., Xi, S., Jiang, H., et al.: Resource allocation and offloading strategy for UAV-assisted LEO satellite edge computing. Drones **7**(6), 383 (2023)
48. Han, D., Ye, Q., Peng, H., et al.: Two-timescale learning-based task offloading for remote IoT in integrated satellite-terrestrial networks. IEEE Internet Things J. **10**(12), 10131–10145 (2023)
49. Luo, X., Chen, H.H., Guo, Q.: Semantic communications: overview, open issues, and future research directions. IEEE Wirel. Commun. **29**(1), 210–219 (2022)
50. Dai, J., Zhang, P., Niu, K., et al.: Communication beyond transmitting bits: semantics-guided source and channel coding. IEEE Wirel. Commun. **4**, 170–177 (2023)

Energy-Constrained Model Pruning for Efficient In-Orbit Object Detection in Optical Remote Sensing Images

Shaohua Qiu[1,3](✉) ⓘ, Du Chen[2], Xinghua Xu[1], and Jia Liu[2]

[1] National Key Laboratory of Electromagnetic Energy, Naval University of Engineering,
Wuhan 430033, China
qiush125@163.com

[2] School of Computer Science, China University of Geosciences, Wuhan 430074, China

[3] East Lake Laboratory, Wuhan 430202, China

Abstract. Efficient object detection from optical remote sensing (RS) images has always been an important interpretation task for in-orbit RS applications. In recent years, convolutional neural networks have been widely used for object detection with significantly improved detection accuracy. However, the large detection models pose great challenges for the computing, memory and energy supply of resource-constrained in-orbit platforms. In this paper, we propose an efficient in-orbit object detection method with low memory, computation and energy requirements. The proposed method first integrates the compact modules of GhostNet into the detector and further performs the L1-norm based filter pruning to significantly reduce model size and computational complexity. Besides, we propose to use energy as a key metric in filter pruning, and present a novel energy-guided layer-wise pruning rate estimation method so as to achieve energy-efficient object detection. Comprehensive experiments have shown the effectiveness of the proposed method in terms of model size, computational complexity, latency and energy consumption, while maintaining comparable detection accuracy.

Keywords: In-Orbit Object Detection · Optical Remote Sensing Images · Constrained Resources · Lightweight CNN · Filter Pruning

1 Introduction

Object detection from optical remote sensing (RS) imagery tries to localize geospatial objects of interest, e.g. airplanes, ships, etc., with their corresponding categories. It is a crucial task of image interpretation, and has been widely applied in a variety of applications such as environmental monitoring, disaster reduction. In recent years, object detectors with deep learning, especially convolutional neural networks (CNN), have achieved success in RS community. Methods have been developed considering the specific issues of RS scenes, e.g. rotation, multi-scale, complex background, tiny objects [1–3]. They generally design large models to enhance feature extraction capacity, thereby improving accuracy. It results in an increase in model size, computational complexity and energy consumption, which poses huge challenges for their implementations on resource-constrained platforms [4].

Traditional RS applications typically follow the workflow of "acquisition, delivery, and processing", which can lead to delay for time-sensitive object detection tasks. In recent years, with the development of sensor technologies, the number of RS devices and edge infrastructures (satellite constellation, etc.) keeps increasing. Intelligent RS satellites such as LuoJia3 [5], have been launched successively. These platforms are generally equipped with real-time edge computing architecture, which enables the in-orbit RS data processing and intelligent inference. To deploy the deep learning based RS image object detection models in the edge platforms closer to the data source can effectively reduce redundant data transfer and response latency [6]. However, these edge platforms are with limited computing, storage and power supply resources.

On the other hand, studies have found that redundant information exist in large models, and not all parameters and structures contribute to the high discriminability [7, 8]. Therefore, compact networks have been designed, such as MobileNetV1 [9], MobileNetV2 [10] and ShuffleNet [11], which rely on hand-crafted features and need domain experts to obtain a design solution with the tradeoff of model size, efficiency and accuracy. Recently, several studies, e.g. MobileNetV3 [12] and FBNet [13], can derive compact networks by neural architecture search (NAS) rather than manually designing networks [14]. NAS is a sub-domain of auto machine learning (ML) that requires extensive computing resources [15].

In addition, researchers have tried to compress and accelerate large models while maintaining high accuracy. The common ways can be divided into model pruning, low-rank factorization, parameter quantization, and knowledge distillation [4]. Low-rank factorization attempts to factorize large weight matrix/tensor into smaller ones layer by layer, which normally involves in convolutional layer and fully-connected layer, while its decomposition operation is computationally expensive [7]. Parameter quantization performs the low-bit representation for weights which are generally 32-bit floating-point numbers, normally leading to an accuracy decrease [16]. Knowledge distillation is to train a lightweight student model from a larger teacher model while maintaining its generalization ability [8]. Generally, its performance is sensitive to the network structures, and requires training from scratch. Parameter pruning attempts to reduce redundant parameters that contribute little to the performance, and has been widely used to reduce network complexity and address the over-fitting issue [7]. In parameter pruning, different granularity can be performed, e.g., neuron-level pruning, filter-level pruning and layer-level pruning.

Specifically, to meet the requirements of resource-constrained RS scenarios, e.g. onboard satellite platforms, there are researches towards lightweight object detectors for RS imagery. Miao et al. replaced the shallow convolutional layers in ResNet-50 with Ghost modules to achieve lightweight ship detection from RS images [17]. To adapt to the edge computing devices, Pang et al. proposed a lightweight fine-grained object recognition network SOCNet with flat feature extraction backbones and depth-wise separable convolutions to accelerate inference [18]. Some researches attempt to achieve near real-time RS object detection with lightweight CNN as feature extraction backbone in one-stage detectors, e.g., SSD with MobileNet [19], R2-CNN with its backbone Tiny-Net [20], RetinaNet with Ghost module [17], YOLO with MobileNet [21]. Besides, a few studies explore to reduce model size, computational complexity, and latency of RS object detection through model compression, including network pruning [22–24], low-rank factorization [25], model quantization [26, 27], knowledge distillation [28–30], as well as their combinations [31].

In addition, edge devices aboard satellites are also with constrained energy supply [32]. However, the previous studies mainly focus on the metrics of computational complexity, model size and runtime. To address this issue, this paper introduces the energy consumption as a major factor into model pruning, so as to reduce the model parameters, computational complexity, as well as inference energy consumption of in-orbit object detection model. The major contributions are as follows:

(1) A lightweight detection model has been first implemented based on YOLOv5, combining the compact CNN modules from GhostNet and filter pruning. It can reduce the model size, computational complexity and latency significantly.
(2) We use energy constraint as a metric in filter pruning, and propose a novel and automatic energy-guided layer-wise pruning rate estimation method, so as to achieve energy-efficient detection from RS imagery.
(3) Comprehensive experiments on the dataset DOTA-v2.0 show the effectiveness of the proposed method in detection accuracy, model size, computational complexity, latency, as well as energy consumption.

2   The Proposed Method

2.1   Overview

The overview of the proposed method is shown in Fig. 1. Given the fact that the in-plane directions of objects in RS images are arbitrary, an oriented object detection method is implemented firstly by using the rotated bounding box representation, rotation-robust Intersection-over-Union (IoU) and an angle-sensitive loss function. It is utilized as the baseline for the subsequent improvements. After that, the lightweight Ghost convolution is incorporated to reduce parameters and computation in backbone. Subsequently, we adopt the classical filter selection criterion of "smaller norm - less importance", and perform the structured filter pruning based on L1 norm to remove redundant filters, thus further increasing the sparsity of the detection model. Finally, taking the energy consumption of model inference into consideration, a novel energy-guided layer-wise pruning rate estimation method is proposed. It is performed with the L1-norm based filter pruning to meet the energy consumption requirement of RS object detection. The details will be described in details in the following subsections.

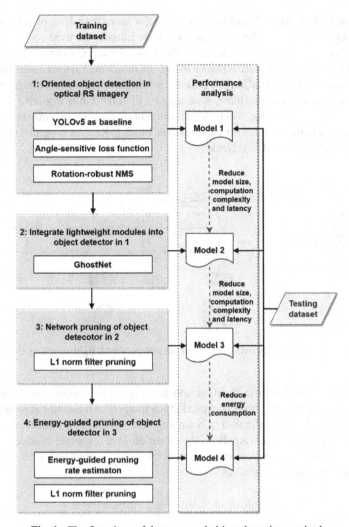

Fig. 1. The flowchart of the proposed object detection method.

2.2 Lightweight Oriented Object Detection Network

In this paper, the one-stage YOLOv5l is used as the baseline. It consists of backbone, neck and head. The backbone extracts features from input images, and plays a crucial role in detection performance. It adopts a new CSPDarknet as backbone, which is based on cross stage partial networks (CSP) and Focus structure. The neck fuses the extracted features, and a new CSP path aggregation network structure (CSP-PAN) have been employed. The head performs the final detection, and is consistent with the previous YOLOv3 and YOLOv4.

Since the CSPDarknet53 backbone accounts for more than 50% of the parameters and computations in the detection model, we have initially focused on slimming this heavy backbone by incorporating lightweight modules from GhostNet. The improved lightweight network architecture is shown in Fig. 2. GhostNet [33] uses a novel plug-and-play Ghost module, i.e. GhostConv to generate informative ghosts features from a series of linear transformations. C3Ghost is composed of three GhostBottleNeck modules stacked in a sequential order, where Ghost bottleneck consists of two stacked Ghost modules and shortcuts similar to the basic residual block in ResNet. Note that the compact modules are also integrated into Neck.

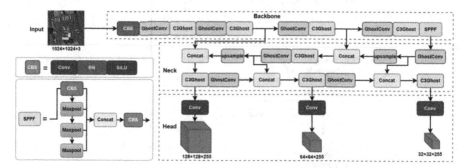

Fig. 2. The baseline detection network architecture with lightweight module.

The detector's output, loss function and non-maximum suppression (NMS) have been adapted to the oriented objects existed in RS imagery. An additional rotation angle parameter is added to the definition of bounding box, which better reflects the size and aspect ratio of objects. Accordingly, a classification loss for angle is added in Eq. 1. The rotation-robust intersection over union (RIoU) [34] is used for NMS to filter out duplicate detection bounding boxes.

$$L_{total} = \lambda_{box}L_{box} + \lambda_{obj}L_{obj} + \lambda_{cls}L_{cls} + \lambda_{\theta}L_{\theta} \tag{1}$$

In Eq. 1, L_{box} is the localization loss for bounding box regression and uses complete intersection over union (CIoU) loss $L_{box} = 1 - \text{CIoU}$ in [35]; L_{cls} and L_{obj} are the classification loss and confidence loss respectively, both adopt binary cross entropy (BCE) loss; L_{θ} is the added angle classification loss shown in Eq. 2, where y is the ground truth and $\hat{y}$ is the predicted value, and we take it as a classification task with 180 values, i.e. $\theta \in [-90, 90)$ and is an integer. λ_{box}, λ_{cls}, λ_{obj} and λ_{θ} represent the proportion of L_{box}, L_{cls}, L_{obj} and L_{θ} in L_{total}.

$$L_{\theta} = -\sum_{i=1}^{N} y^{(i)} \log \hat{y}^{(i)} + (1 - y^{(i)}) \log(1 - \hat{y}^{(i)}) \tag{2}$$

2.3   Filter Pruning of RS Object Detection with L1-Norm

After integrating the lightweight CNN modules from GhostNet into the detector, the parameters and computations have been reduced significantly, but still contain much redundancy, which can be compressed further. Model pruning removes redundant parameters according to the designed evaluation criteria for network parameters based on pre-trained large models. It can be divided into unstructured pruning and structured pruning, according to the pruned components. The fine-grained unstructured pruning such as weight pruning, can greatly reduce model size and computation by removing redundant weights while maintaining accuracy, however requires special hardware and runtime libraries to accelerate the resulting unstructured sparsity [36]. In contrast, the coarse-grained structured pruning is preferred to compress and accelerate models under existing platforms, where filter pruning is a widely adopted strategy [37, 38].

Following a classical criterion "smaller-norm-less-important", this paper applies a filter pruning process with smaller L1-norm values in the overall detection network derived from the previous subsections. The corresponding pseudo-code is given in Algorithm 1. The filters with smaller L1-norm values than a threshold are identified as unimportant and pruned. The threshold is determined by a given pruning rate p. The L1-norm values of all filters are sorted in ascending order, and the threshold is then the L1-norm value of the $round(N \times p)$ filters, where N is the filter numbers of a layer. In this paper, the pruning rate p can vary across layers and is set as 0.5, namely 50% filters will be pruned for all convolutional layers. The pruned model are then fine-tuned for performance recovery.

Algorithm 1. Filter pruning with L1-Norm for the detection model.

Input: Filter of each layer x , Number of layers n , Pruning rate p

---

Output: Pruned detector

---

   for $i = 0$ to $n-1$ **do**
      filter $\leftarrow x[i]$;
      $N = filter.shape[0]$;
      for $j = 0$ to $N-1$ **do**
         $array[j] = \text{L1}(filter[j])$;
      end
      $copyArray = \text{copy}(array)$;
      $\text{sortByL1}(array)$;
      $threshold = array[Round(p \times N) - 1]$;
      for $j = 0$ to $N - 1$ **do**
         if $copyArray[j] < threshold$ **then**
            $\text{pruning}(x[i], copyArray, j)$;
         end
      end
   end

2.4 Energy-Guided Layer-Wise Pruning Rate Estimation

Energy consumption is an indicator that does not simply linearly decrease along with the model size and computations. It is important but often ignored in current research of lightweight RS object detection. Therefore, this subsection presents a simple yet practical energy-guided layer-wise pruning rate estimation method to achieve energy-efficient object detection in RS images.

The energy consumption is first measured and calculated for the model inference in Subsect. 2.2 and Subsect. 2.3. With a fixed size (3 × 1024 × 1024) inputs, the start time T_0 and end time T_n have been recorded. The time T_i and the corresponding instantaneous power of graphics processing unit (GPU) P_i at regular intervals are measured using the command *nvidia-smi* provided by the official NVIDIA plug-in. The energy consumption is measured and approximated for model inference as shown Eq. 3. In this work, and the energy consumption of each model is measured 10 times to obtain the average value.

$$W = \sum_{i=1}^{n} (T_i - T_{i-1}) \times P_i \tag{3}$$

In Subsect. 2.3, we apply a fixed pruning rate of 50%. However, different layers contribute varied energy consumption. Therefore, we have constructed a dataset with 5000 records which reflects the relations of parameters and computations with energy consumption through randomly generating the sparsity of layers and measuring the corresponding energy consumption. The records in the dataset indicate that the energy consumption maintains an approximate linear correlation with the computations in floating point of operations (FLOPs) for the detection model in previous subsections. This observed correlation does not hold between energy consumption and parameters. It provides the foundation for the following pruning rate estimation method.

The corresponding pseudo-code for pruning rate estimation is in Algorithm 2. The energy consumption to be reduced for model inference has been dispersed to layers, which can be defined by Eq. 4. W_t is the defined target energy consumption, W_o is the original energy consumption, and the energy consumption reduced by pruning is $W_s = W_o - W_t$.

$$W_i = W_s \times FLOPs_i \left/ \sum_{i=1}^{n} FLOPs_i \right. \tag{4}$$

Algorithm 2. Energy-guided pruning rate estimation.

Input: Model x, reduced energy consumption W_s, number of layers n, computations of a layer $FLOPs$, tolerance $error$

Output: Estimated pruning rate arr

---

for $i = 0$ **to** $n - 1$ **do**
    $W[i] = Ws \times FLOPs[i]/\text{sum}(FLOPs)$;
end
sortByFlops(W);
for $i = 0$ **to** $n-1$ **do**
    for $j = 0.25$ **to** 0.75 **do**
        $w = \text{getEnergy}(x, j)$;
        $k = \text{getLayerIndex}(W)$;
        if $|w - W[i]| < error$ **then**
            $arr[k] = j$;
            break;
        end
    end
end

---

3 Experimental Results and Analysis

3.1 Dataset, Experimental Settings and Evaluation Metrics

A large-scale aerial RS image dataset for object detection, DOTA-v2.0, is adopted for performance evaluation. Its images are collected from Google Earth, GF-2 Satellite and aerial images. Image sizes range from 800×800 to $20,000 \times 20,000$ pixels and each image contains objects exhibiting a wide variety of scales, orientations, and shapes. There are 18 common categories with 11,268 images and 1,793,658 instances. The categories in DOTA-v2.0 include plane (PL), ship (SP), storage tank (ST), baseball diamond (BD), tennis court (TC), basketball court (BC), ground track field (GTF), harbor (HB), bridge (BG), large vehicle (LV), small vehicle (SV), helicopter (HC), roundabout (RA), soccer ball field (SBF), swimming pool (SWP), container crane (CC), airport (AP) and helipad (HP).

The images have been cropped into a fixed size of 1024×1024 with an overlapping width of 200 pixels, according to GPU memory. The data augmentation, e.g. rotating, mirroring, and adding noise etc. has been performed. For the rotated annotations, we convert the original 8-d.o.f.-parameter label into 5-parameter one with rotating angle.

Experiments are performed on a server with an Intel Xeon E5-2680v4 CPU, 64GB random access memory (RAM) and an NVIDIA RTX 3090 GPU. The system runs Ubuntu 20.04 LTS with CUDA 11.1, cuDNN 8.0.5, Pytorch 1.8.0, and torchvision 0.9.0. Models are trained with a batch size of 16 by stochastic gradient descent (SGD) with an initial learning rate of 0.01 and momentum of 0.937 for 250 epochs. The learning rate is adjusted using the cosine annealing algorithm.

The detection accuracy is evaluated in mAP. The detection efficiency is evaluated with model sizes (MB), computational complexity (FLOPs), frame per second (FPS), and energy consumption (J). FLOPs refers to the number of floating point operations performed for model inference, and is a common metric to measure the computational complexity. FPS refers to the number of images inferred per second, and the time of each image inferred is *Latency*. The energy consumption can be obtained by Eq. 3.

3.2 Detection Results and Accuracy

The experimental results of detection accuracy including AP for each category and mAP (%) are in Table 1. YOLOv5l, YOLOv5l-G, YOLOv5l-GP and YOLOv5l-GPE stand for the adapted YOLOv5 for oriented object detection, the lightweight YOLOv5 with compact modules, the model using filter pruning with L1-norm (fixed pruning rate of 0.5), and the energy-guided model (energy constraints ratio of 0.5).

Table 1. Detection accuracy (%) on the DOTA-v2.0 dataset.

Class	Model			
	YOLOv5l	YOLOv5l-G	YOLOv5l-GP	YOLOv5l-GPE
PL	**97.2**	96.9	95.9	96.0
BD	78.2	**79.0**	76.3	76.8
BG	**57.0**	56.6	49.9	51.1
GTF	71.2	72.1	72.4	**73.1**
SV	**73.8**	71.2	67.6	67.6
LV	**84.9**	83.7	82.7	82.5
SP	**95.1**	93.8	92.4	92.5
TC	**96.7**	96.3	95.7	95.9
BC	**79.2**	77.9	75.0	76.9
ST	**73.2**	69.3	66.6	70.1
SBF	**54.3**	49.6	52.2	50.7
RA	60.9	58.1	63.9	**66.6**
HB	**82.5**	81.8	80.7	81.3
SWP	70.9	**72.5**	65.8	68.7
HC	74.6	71.8	62.5	**74.9**
CC	**3.5**	0.9	1.4	1.0
AP	28.0	64.7	**66.8**	51.4
HP	**50.9**	50.6	33.5	50.0
mAP	68.4	**69.3**	66.7	68.2

Table 1 shows that all detection models perform well for the categories with more training data and distinct geometric features, e.g., plane, ship and tennis court, while the accuracy varies for the categories with less training data, e.g., helipad, container crane and airport. Specifically, YOLOv5l achieves a mAP of 68.4%, with the highest accuracy in AP for the majority of categories. YOLOv5l-G without pruning has a slightly higher mAP of 69.3%, and YOLOv5l-GP which prunes YOLOv5l-G using a fixed pruning rate of 0.5 obtains an accuracy of 66.7%. The mAP of the final energy-efficient YOLOv5l-GPE is 68.2%, only with a slight accuracy loss of 1.1% and 0.2% compared with the unpruned YOLOv5l-G and baseline YOLOv5l. Some detection examples on the benchmark dataset DOTA-v2.0 are shown in Fig. 3.

Fig. 3. Detection examples of the energy-efficient detection model on the DOTA-v2.0 dataset.

3.3 Model Size and Computational Complexity

In addition to detection accuracy, the parameters, model size (MB) and computational complexity (GFLOPs) are presented in Table 2 to evaluate efficiency performance. It can be observed that the model size of YOLOv5l-G decreases from 71.7 MB to 32.8 MB when using float16. YOLOv5l-GP exploits the redundancy of detection models, and further significantly reduces the model size to 9.61 MB, which account for only 13.40% of the baseline model YOLOv5l, and 29.30% of YOLOv5l-G. When setting an energy consumption constraint ratio of 0.5, YOLOv5l-GPE reduces to 12.4 MB with a compression ratio of 5.78× and 2.65×, compared to YOLOv5l and YOLOv5l-G respectively. Its model size compression efficiency decrease, but is with increased accuracy compared with the YOLOv5l-GP.

Computational complexity is measured in GFLOPs, which does not corresponds to parameters directly. Generally, convolutional layers contain less parameters, but are computationally intensive, while the full connection layers are on the contrary. The computation complexities of YOLOv5l-G, YOLOv5l-GP, YOLOv5l-GPE decrease to

Table 2. Model size, computational complexity, latency, FPS and energy consumption.

Metric	Model			
	YOLOv5l	YOLOv5l-G	YOLOv5l-GP	YOLOv5l-GPE
mAP (%)	68.4	69.3	66.7	68.2
Model size (MB)	71.7	32.8	9.61	12.4
Computational Complexity (GFLOPs)	217.9	93.1	27.0	29.6
Latency (ms)	24.5	22.3	20.4	20.3
FPS	40.8	44.8	49.0	49.3
Energy Consumption (J)	1158.3	1097.4	590.5	580.2

93.1 GFLOPs, 27.0 GFLOPs and 29.6 GFLOPs respectively, accounting for 42.73%, 12.39% and 13.58% of the baseline YOLOv5l. The compression ratio of FLOPs is slightly significant than that of the model size.

The detection latency and FPS achieve only a slight improvement, and the highest speedup is 1.21×, as shown in Table 2. It can be observed that FLOPs cannot be used directly as an indicator of inference speed, i.e. low FLOPs does not correspond to high FPS. The inference speed is influenced by many factors, e.g. memory access cost, parallelism and platforms. It should be noted that the efficiency improvement due to the parallelism on the corresponding accelerators is not the focus of this paper, so the latency and FPS have not be tuned.

3.4   Energy Consumption

In this subsection, we further explore the energy consumption for detection. Table 2 shows the energy consumption results for the YOLOv5l, YOLOv5l-G, YOLOv5l-GP (a fixed pruning rate of 0.5), and YOLOv5l-GPE (an energy constraints ratio of 0.5). The energy consumption results (J) in Table 2 refer to the model inference with an image of 3 × 1024 × 1024 for 100 times, and the measurements have been performed 10 times for the average. It can be seen that, although the energy-guided YOLOv5l-GPE has slightly higher model size and computational complexity, it consumes less energy of 580.2 J than YOLOv5l-GP.

We vary the energy constraint configurations with 0.4, 0.5 and 0.6 for YOLOv5l-GPE. The setting of 0.4 represents that a decline by 40% in energy consumption is desired. For a detection model, the same pruning sparsity on different layers may lead to varied influences on the total energy consumption. The layer-wise pruning rates are first estimated according to Subsect. 2.4. The corresponding results are given in Fig. 4. Note that the layers with an estimated pruning rate of 0 involve no convolutions. The estimated pruning rates vary, and are in 0.25–0.69 (setting of 0.4), 0.25–0.74 (setting of 0.5), and 0.32–0.74 (setting of 0.6), respectively. The maximum pruning rate for each layer is 0.75 for maintaining detection accuracy.

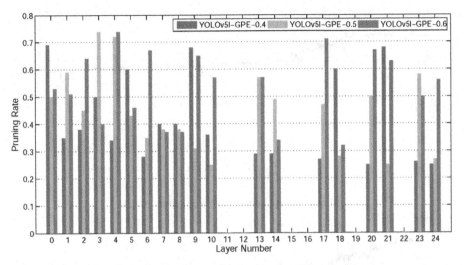

Fig. 4. The layer-wise pruning rates under different settings.

In addition, we present the consumption constraint variations with the pruning rate adjustment in Fig. 5. With the continuous adjustment of sparsity along the iterations, the energy consumption of model decreases in general, and the final result approaches the desired constraint. The fluctuations in the energy consumption mainly appear in the process of pruning rate search within a certain layer, and the iteration number with the final estimated results are marked with the layer number. The slight fluctuations across layers are mainly due to that the pruning rates to meet the requirements are unavailable, as well as deviations exist in the energy consumption measurements. Figure 5 also shows the layer order of pruning rate estimation, namely, from the layer with the highest proportion of FLOPs to that with the lowest as depicted in Subsect. 2.4. The orders are consistent with different energy consumption constraints.

The energy consumption results under varied constraint settings are also presented in Table 2. YOLOv5l-G is the baseline for the settings of Energy-0.4, Energy-0.5 and Energy-0.6. The YOLOv5l-GPE with a higher energy constraint achieves a lower model size, computational complexity and energy consumption. There is a decline in detection accuracy. With the corresponding energy consumption constraint, the actual energy consumption reduction is 448.4J, 517.2J and 548.7J, which corresponds to 40.86%, 47.13% and 50% of YOLOv5l-G. It indicates that the energy consumption decline is approaching the target constraint 40%, 50% and 60%. However, higher energy consumption constraint setting makes this process more difficult.

46   S. Qiu et al.

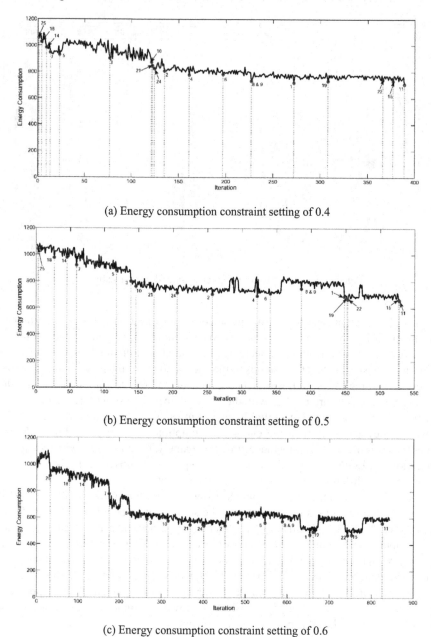

(a) Energy consumption constraint setting of 0.4

(b) Energy consumption constraint setting of 0.5

(c) Energy consumption constraint setting of 0.6

Fig. 5. The consumption constraint variations during the pruning rate adjustment.

4 Conclusion

This paper proposes a novel energy-constrained model pruning method for efficient in-orbit object detection in optical remote sensing images, thus reducing parameters, computation complexity, and energy consumption with competitive detection accuracy. It is achieved by lightweight CNN and filter pruning supported with a novel energy-guided layer-wise pruning rate estimation method. Comprehensive experiments show that the compact modules and a simple L1-norm based filter pruning can achieve a minimal model size of 9.61MB (7.46×), computational complexity of 27.0 GFLOPs (8.07×) compared with the baseline YOLOv5l, with an accuracy loss of 1.7% in mAP. Considering the energy consumption, the layer-wise pruning rates are estimated, and the resulting YOLOv5l-GPE model can lead to a lower energy consumption (40.86%–50%) approaching to the constraint settings.

Acknowledgments. This work is supported by National Natural Science Foundation of China under Grant 41901376, Hubei Provincial Natural Science Foundation of China under Grant 2022CFB989, and Foundation for the National Key Laboratory under Grant 6142217210503, and China University of Geosciences (Wuhan) Teaching Laboratory Open Foundation SKJ2022230.

Disclosure of Interests. The authors have no competing interests to declare that are relevant to the content of this article.

References

1. Dong, Z., Wang, M., Wang, Y., Zhu, Y., Zhang, Z.: Object detection in high resolution remote sensing imagery based on convolutional neural networks with suitable object scale features. IEEE Trans. Geosci. Remote Sens. **58**, 2104–2114 (2020)
2. Yu, D., Ji, S.: A new spatial-oriented object detection framework for remote sensing images. IEEE Trans. Geosci. Remote Sens. **60**, 1–16 (2022)
3. Xu, Z., Rao, M.: Multiscale information fusion-based deep learning framework for campus vehicle detection. Int. J. Image Data Fusion **12**, 83–97 (2021)
4. Liu, J., Xiang, J., Jin, Y., Liu, R., Yan, J., Wang, L.: Boost precision agriculture with unmanned aerial vehicle remote sensing and edge intelligence: a survey. Remote Sens. **13** (2021)
5. Zhang, Z., Qu, Z., Liu, S., Li, D., Cao, J., Xie, G.: Expandable on-board real-time edge computing architecture for Luojia3 intelligent remote sensing satellite. Remote Sens. **14**, 3596 (2022)
6. Ren, F., Li, Y., Zheng, Z., Yan, H., Du, Q.: Online emergency mapping based on disaster scenario and data integration. Int. J. Image Data Fusion **12**, 282–300 (2021)
7. Cheng, Y., Wang, D., Zhou, P., Zhang, T.: Model compression and acceleration for deep neural networks: the principles, progress, and challenges. IEEE Signal Process. Mag. **35**, 126–136 (2018)
8. Choudhary, T., Mishra, V., Goswami, A., Sarangapani, J.: A comprehensive survey on model compression and acceleration. Artif. Intell. Rev. **53**, 5113–5155 (2020)
9. Howard, A.G., et al.: MobileNets: efficient convolutional neural networks for mobile vision applications. arXiv:1704.04861 (2017)
10. Sandler, M., Howard, A., Zhu, M.L., Zhmoginov, A., Chen, L.C.: MobileNetV2: inverted residuals and linear bottlenecks (2018)

11. Zhang, X., Zhou, X.Y., Lin, M.X., Sun, R.: ShuffleNet: an extremely efficient convolutional neural network for mobile devices (2018)
12. Howard, A., et al.: Searching for MobileNetV3, pp. 1314–1324 (2019)
13. Wu, B., et al.: FBNet: hardware-aware efficient ConvNet design via differentiable neural architecture search (2019)
14. Elsken, T., Metzen, J.H., Hutter, F.: Neural architecture search: a survey. J. Mach. Learn. Res. **20** (2019)
15. He, X., Zhao, K.Y., Chu, X.W.: AutoML: a survey of the state-of-the-art. Knowl.-Based Syst. **212** (2021)
16. Li, Z.S., et al.: A compression pipeline for one-stage object detection model. J. Real-Time Image Proc. **18**, 1949–1962 (2021)
17. Miao, T., et al.: An improved lightweight RetinaNet for ship detection in SAR images. IEEE J. Sel. Top. Appl. Earth Obs. Remote Sens. **15**, 4667–4679 (2022)
18. Pang, Y., Zhang, Y., Wang, Y., Wei, X., Chen, B.: SOCNet: a lightweight and fine-grained object recognition network for satellite on-orbit computing. IEEE Trans. Geosci. Remote Sens. **60**, 1–13 (2022)
19. Li, L., Zhang, S., Wu, J.: Efficient object detection framework and hardware architecture for remote sensing images. Remote Sens. **11**, 2376 (2019)
20. Pang, J., Li, C., Shi, J., Xu, Z., Feng, H.: R2-CNN: fast tiny object detection in large-scale remote sensing images. IEEE Trans. Geosci. Remote Sens. **57**, 5512–5524 (2019)
21. Hong, Q., et al.: A lightweight model for wheat ear fusarium head blight detection based on RGB images. Remote Sens. **14**, 3481 (2022)
22. Deng, G.C., et al.: A low coupling and lightweight algorithm for ship detection in optical remote sensing images. IEEE Geosci. Remote Sens. Lett. **19** (2022)
23. Zhong, C.L., Mu, X.D., He, X.C., Wang, J.X., Zhu, M.: SAR target image classification based on transfer learning and model compression. IEEE Geosci. Remote Sens. Lett. **16**, 412–416 (2019)
24. Ma, X.J., Ji, K.F., Xiong, B.L., Zhang, L.B., Feng, S.J., Kuang, G.Y.: Light-YOLOv4: an edge-device oriented target detection method for remote sensing images. IEEE J. Sel. Top. Appl. Earth Obs. Remote Sens. **14**, 10808–10820 (2021)
25. Xue, W., Qi, J.H., Shao, G.Q., Xiao, Z.X., Zhang, Y., Zhong, P.: Low-rank approximation and multiple sparse constraint modeling for infrared low-flying fixed-wing UAV detection. IEEE J. Sel. Top. Appl. Earth Obs. Remote Sens. **14**, 4150–4166 (2021)
26. Wei, X., Liu, W.C., Chen, L., Ma, L., Chen, H., Zhuang, Y.: FPGA-based hybrid-type implementation of quantized neural networks for remote sensing applications. Sensors **19** (2019)
27. Zhang, R.Y., Jiang, X.J., An, J.S., Cui, T.S.: Data-free low-bit quantization for remote sensing object detection. IEEE Geosci. Remote Sens. Lett. **19** (2022)
28. Chen, J.Z., Wang, S.H., Chen, L., Cai, H.B., Qian, Y.T.: Incremental detection of remote sensing objects with feature pyramid and knowledge distillation. IEEE Trans. Geosci. Remote Sens. **60** (2022)
29. Yang, Y.R., Sun, X., Diao, W.H., Yin, D.S., Yang, Z.J., Li, X.M.: Statistical sample selection and multivariate knowledge mining for lightweight detectors in remote sensing imagery. IEEE Trans. Geosci. Remote Sens. **60** (2022)
30. Zhang, Y.D., Yan, Z.Y., Sun, X., Diao, W.H., Fu, K., Wang, L.: Learning efficient and accurate detectors with dynamic knowledge distillation in remote sensing imagery. IEEE Trans. Geosci. Remote Sens. **60** (2022)
31. Zhang, F., Liu, Y.B., Zhou, Y.S., Yin, Q., Li, H.C.: A lossless lightweight CNN design for SAR target recognition. Remote Sens. Lett. **11**, 485–494 (2020)
32. Guo, C., Wang, X., Zhong, Z., Song, J.: Research advance on neural network lightweight for energy optimization. Chin. J. Comput. **46**, 85–102 (2023)

33. Han, K., Wang, Y., Tian, Q., Guo, J., Xu, C., Xu, C.: GhostNet: more features from cheap operations. In: Proceedings of the IEEE/CVF Conference on Computer Vision and Pattern Recognition (CVPR), pp. 1580–1589 (2020)
34. Zheng, Y., Zhang, D., Xie, S., Lu, J., Zhou, J.: Rotation-robust intersection over union for 3D object detection. In: Vedaldi, A., Bischof, H., Brox, T., Frahm, J.-M. (eds.) Computer Vision – ECCV 2020. LNCS, vol. 12365, pp. 464–480. Springer, Cham (2020). https://doi.org/10.1007/978-3-030-58565-5_28
35. Zheng, Z., Wang, P., Liu, W., Li, J., Ye, R., Ren, D.: Distance-IoU loss: faster and better learning for bounding box regression. In: AAAI Conference on Artificial Intelligence, vol. 34, pp. 12993–13000 (2020)
36. Xie, X., Lin, J., Wang, Z., Wei, J.: An efficient and flexible accelerator design for sparse convolutional neural networks. IEEE Trans. Circuits Syst. I-Regul. Pap. **68**, 2936–2949 (2021)
37. Luo, J.-H., Wu, J., Lin, W.: ThiNet: a filter level pruning method for deep neural network compression. In: 2017 IEEE International Conference on Computer Vision (ICCV), pp. 5068–5076. IEEE (2017)
38. He, Y., Liu, P., Wang, Z., Hu, Z., Yang, Y.: Filter pruning via geometric median for deep convolutional neural networks acceleration. In: 2019 IEEE/CVF Conference on Computer Vision and Pattern Recognition (CVPR 2019), pp. 4335–4344 (2019)

Cooperative Beam Hopping for LEO Constellation Network

Chaoyu Ren[1,2], Feng Tian[2], Wenqian Wang[2,3], and Huijie Liu[2(✉)]

[1] Shanghaitech University, Shanghai 201210, China
renchy2022@shanghaitech.edu.cn
[2] Innovation Academy for Microsatellite of Chinese Academy of Science,
Shanghai 201306, China
liuhj@microsate.com
[3] University of Chinese Academy of Science, Beijing, China

Abstract. LEO constellation network has become a promising approach for global communications, due to its global coverage, lower latency, and flexible construction. However, because of the limited coverage of a single satellite and the high-speed space-ground relative movement, how to leverage the precise satellite resources to satisfy the user terminals' requests is an important issue. Beam hopping technique is an efficient way to handle this issue by scheduling the onboard resource according to the distribution of terminals' requests. In this paper we propose a cooperative beam hopping for LEO constellation network, where multiple satellites provide service for the user terminals in the multi-coverage areas. The problem is innovatively modeled as a LASSO problem considering its sparsity, and machine learning methods are introduced to optimize it. Compared with the fixed beam system, our proposed scheme achieves better performance in terms of resource utilization and average user satisfaction rate.

Keywords: LEO constellation network · Beam hopping · Resource allocation

1 Introduction

Mega LEO constellation communication networks have been an indispensable part of the global network because of the rapid development of advanced satellite manufacturing technology, launch technology, and network technology [9]. LEO constellation has reached unprecedented scale because of its lower cost and smaller size. The largest satellite communications operator SpaceX company has launched 4983 LEO satellites as of August 22, 2023, and 12,000 'StarLink' satellites are expected to be deployed by 2024. Such huge scale will achieve global coverage which can help bridge the "Digital divide" by providing affordable and reliable connectivity services in areas where terrestrial networks are difficult to build [2]. The low latency and small propagation loss of LEO satellites have

© The Author(s), under exclusive license to Springer Nature Singapore Pte Ltd. 2024
Q. Yu (Ed.): SINC 2023, CCIS 2057, pp. 50–59, 2024.
https://doi.org/10.1007/978-981-97-1568-8_5

also earned them ability to provide better global broadband Internet and Internet of Things services [7]. Integrated with ground communication network, LEO constellation network can directly realize 3D global coverage different from traditional 2D 'population coverage' on the surface.

However, due to the uneven distribution of terrestrial user terminals and limited resource on board, the spectrum utilization and service capability of LEO constellation network is lower than terrestrial network [10]. Beam hopping technique in which only a subset of beams is illuminated at a given time for multi-beam satellites [1] is introduced to address this imbalance between system resources and traffic demands in coverage area. In the few past year, beam hopping for single satellite was an appealing issue. An efficient beam hopping based system for single LEO satellite was proposed to allocate onboard resources flexibly according to traffic demand distribution [8]. In order to avoid intra-satellite and inter-satellite interference, a multi-satellite beam hopping algorithm based on load balancing and interference avoidance was proposed for NGSO satellite communication system [6]. Nowadays with the increase of LEO satellites, multi-satellite scenarios have attracted much attention. The determinant point process (DPP) is used to solve LEO dual-satellite dynamic beam hopping problem [5].

But there is few research focused on cooperative beam hopping for multi-satellites especially at high latitude where adjacent satellites always create multiple overlapping areas. By analyzing we find that even near the equator there exist double coverage and at 30-degree latitude triple coverage begins to appear based on our assumption. At higher latitudes, the phenomenon of an area covered by more than three satellites becomes frequent as shown in Fig. 3. Thus, to achieve seamless global coverage and construct mage LEO constellation, it is crucial to research multi-satellites cooperating under multiple coverage.

In this paper, we modeled the scenario as a LASSO problem and a cooperative beam hopping solution is proposed for the first time for multiple coverage. Based on this scheme, we propose a multi-satellite cooperative queue management mechanism to balance the demand and supply. In order to demonstrate advantages of this method, a comparison is made with equivalent fixed-beam systems operating within the same system scenario and conditions. After simulation, our scheme has better performance on user satisfaction and system throughput.

The rest of the paper is organized as follows. Section 2 describes the system model and formulate the multi-satellite beam hopping problem. In Sect. 3, we analyze the scenario we have built and make a comparation with fixed beam condition. Section 4 presents the simulation result. Section 5 concludes the work.

2   System Model and Problem Formulation

2.1   System Model

In this paper, we consider the forward link of the LEO constellation network as shown in Fig. 1. According to the coverage of the satellite beams, the region is divided into different cells covered by virtual beams. Only cells illuminated by real satellite beam can be served and satellites dynamically switch beams in accordance with requests in different cells due to the limited on-board resource. The specific process is displayed in Fig. 2. Once terminal users send request to satellites, multi-beam antenna will adjust the beam illuminated regions according to hopping map formed by resource allocation algorithm based on traffic demand distribution. To be noticed, user requests in multiple coverage for resources are cashed in queues on different satellites [3].

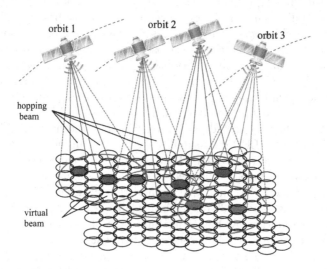

orbit 1   orbit 2   orbit 3

hopping
beam

virtual
beam

Fig. 1. Illustration of cooperative beam hopping of LEO constellation

For simulation, we take a mega LEO constellation into account, where N satellites serve a specific area. The detailed notations and definitions used are summarized in Table 1. To simplify the problem, we assume that each satellite and beam are the same, so suppose that a single multi-beam satellite can provide N_B beams and serve N_V virtual beams.

For one specific satellite, different beams are distributed with time division [4]. Therefore, we divide a period time into time-segments with length T_s where one time-segment is divided into K time slot and the beginning of each time slot denoted as t, $t = 1, 2, \ldots, K$. To balance resource supply and demand, we denote vector $\mathbf{D}^t$ and $\mathbf{X}^t$ to indicate the demand of all covered virtual beam and resource allocated to all beams each satellite at time slot t, where d_j^t and x_k^t respectively representing the demand of jth virtual beam and the capacity of kth beam at time slot t seeing in formula (1). Generally, the lengths of vector $\mathbf{D}^t$ and $\mathbf{X}^t$ are N_V and N_B, respectively, but if we consider some satellites as a cluster, their lengths can be variable.

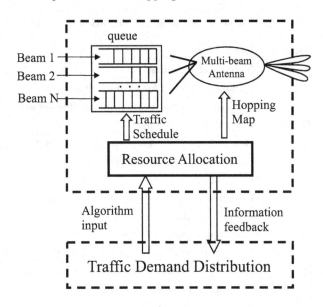

Fig. 2. Illustration of beam-hopping satellite communication system

$$\mathbf{D}^t = \begin{bmatrix} d_1 \\ d_2^t \\ \dots \\ d_n^t \end{bmatrix} \quad \mathbf{X}^t = \begin{bmatrix} x_1 \\ x_2^t \\ \dots \\ x_n^t \end{bmatrix} \tag{1}$$

According to Shannon's channel capacity theorem, we define P_i^t and B_i^t to respectively indicate the power and bandwidth of ith satellite at time slot t. Assume that the total power and bandwidth on-board are marked by P_0 and B_0. So, we can obtain the following constraints:

$$\sum_{i=1}^{N} B_i^t \le B_0 \tag{2}$$

$$\sum_{i=1}^{N} P_i^t \le P_0 \tag{3}$$

Due to the uniform satellite coverage in different region, how to schedule the resources on overlapping area covered by multiple satellites is a remarkable issue. Therefore, we define C_i^t to represent the whole resource supply capacity for beam i at t time slot where N_0 stands for noise and interference as follow:

$$C_i^t = B_i^t \log_2(1 + p_i^t/N_0) \tag{4}$$

The hopping map denoted as a $pN_V \times pN_B$ matrix $\mathbf{M}_i^t$ for i^{th} satellite where $m_{a,b}$ represents whether at time slot t, the bth virtual beam needs to be illuminated by a^{th} beam for a cluster with p satellites. If $m_{a,b} = 1$, the demand is

Table 1. Notation and definition

Notations	Definitions
N	Number of satellites
N_B	Number of beams for each satellite
N_V	Number of virtual beams one satellite served
T_S	Length of time segment
d_i^t	Demand of ith virtual beam at time slot t
x_j^t	Capacity of jth beam at time slot t
$\mathbf{D}^t$	Demand of all coverage at time slot t
$\mathbf{X}^t$	Resource given to all beams at time slot t
C_j^t	Capacity of beam j at time slot t
$\mathbf{M}_i$	The hopping map of ith satellite
B_i^t	Bandwidth to i^{th} satellite at time slot t
P_i^t	Power to i^{th} satellite at time slot t
B_0	The total bandwidth on-board
P_0	The total power on-board
R_i^t	The real traffic for virtual beam i
r_i	Traffic satisfaction rate of virtual beam i
s	Number of satellites covers the same region
v	Number of overlapping virtual beams

totally satisfied, otherwise $m_{a,b} = 0$. Ideally it is expected that all resources are allocated to all users, with no surplus and no waste shown as follow:

$$\mathbf{M}_i^t \cdot \mathbf{X}^t = \mathbf{D}^t \tag{5}$$

The real traffic for i virtual beam is defined as $R_i^t = \min\{c_j^t, d_i^t\}$. We use S_i^t to denote the throughput of terminal users in i virtual beam at t time slot shown as follow:

$$S_i^t = \begin{cases} 0, & m_{a,b} = 0 \\ R_i^t, & m_{a,b} = 1 \end{cases} \tag{6}$$

Another measure to evaluate the system performance is average traffic satisfaction rate defined as follow:

$$r_i = \sum_{t=1}^{K} R_i^t / d_i^t \tag{7}$$

2.2 Problem Formulation

In our cooperative beam hopping system, we pursue higher resource utilizing rate and better services to more users with the limited on-board resources, which means we need to narrow the gap between resources and demand. Considering the overlapping area, we proposed a public resource pool for some satellites to share to avoid resource wasting on public virtual beams. Thus, the resource allocated to each satellite is as less as possible under the condition of satisfying the demand of more users which is represented as:

$$\min \|\mathbf{X}^t\|_1 \tag{8}$$

For single satellite, it is only necessary to ensure the conditions that the capacity each satellite is more than the resource allocated to all beams shown as follow:

$$\sum_{i=1}^{N_B} C_i^t \geq \|\mathbf{X}^t\|_1 \tag{9}$$

But constraint to the overlapping of satellites, public demand should be considered. Suppose that v virtual beams are co-covered by s satellites and traffic in this overlapping area are all stored in queue on-board. Therefore, we only need to guarantee the idle resource of s satellites meet the demand of these virtual beams. The size of matrix $\mathbf{M}_i^t$ are $sN_V \times sN_B$ and the formula of (8) should be looser as:

$$\sum_{i=1}^{sN_B} C_i^t + (s-1) \sum_{i=1}^{v} d_i^t \geq \|\mathbf{X}^t\|_1 \tag{10}$$

In conclude, we construct a LASSO problem to model the scenario as follow:

$$\min \frac{1}{2}\|\mathbf{M}_i^t \cdot \mathbf{X}^t - \mathbf{D}^t\|^2 + \lambda\|\mathbf{X}^t\|_1 \tag{11}$$

For indicators to evaluate the performance of the system shown in formula (6) and (7), the higher maximum of throughput and average traffic satisfaction rate, the better.

3   Subproblem Analysis

In this section we analyze the basic coverage of each satellite and the resource allocation algorithm especially focusing on the satellite queues of multiple overlaps.

Assume that the earth is a standard sphere with radius R. The orbit height of satellites is h and the half-open angle is $\theta/2$. The diagram is shown in Fig. 3. The half satellite coverage corresponds to the central angle of the circle is:

$$\alpha = \pi - \theta/2 - \arcsin((h+R)\sin(\theta/2)/R) \tag{12}$$

The radius of satellite coverage area is αR. As the latitude increases, the coverage area of each satellite in different orbits is more likely to overlap. Assume there are M orbits and LEO satellites are evenly distributed in orbit. The critical overlap latitude ϕ satisfies:

$$2\alpha R \cdot M = 2\pi R \cdot \cos\phi \tag{13}$$

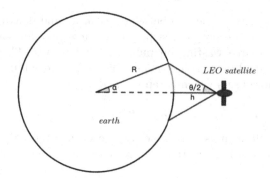

Fig. 3. The illustration of LEO satellite coverage area.

We denote the half-open angle of each beam as $\gamma/2$. Thus, according to formula above, the coverage of each beam can be calculated, and the radius of each virtual beam is denoted as βR.

The problem we have built in (11) is a LASSO problem. Because of many constraints of the problem and the sparsity of matrix $\mathbf{M}_i$, we choose Proximal Gradient (PG) to optimize it which is shown as follow:

Algorithm 1. Proximal Gradient for LASSO

1: Problem:$\min \frac{1}{2}\|\mathbf{M}_i^t \cdot \mathbf{X}^t - \mathbf{D}^t\|^2 + \lambda\|\mathbf{X}^t\|_1$, given $\mathbf{D}^t \in \mathbb{R}^{sN_V}, \mathbf{M}_i^t \in \mathbb{R}^{sN_V \times sN_B}$
2: Input:$\mathbf{x}_0^t \in \mathbb{R}^{sN_B}$ and $L \leq \lambda_{\max}((\mathbf{M}_i^t)^*\mathbf{M}_i^t)$.
3: **repeat**
4:    for $(k = 0, 1, 2, \cdots, K - 1)$ do
5:      $w_k \leftarrow \mathbf{x}_k^t - \frac{1}{L}(\mathbf{M}_i^t)^*(\mathbf{M}_i^t\mathbf{x}_k^t - \mathbf{D}^t)$.
6:      $\mathbf{x}_{k+1}^t \leftarrow \text{soft}(w_k, \lambda/L)$.
7: **until** $\frac{1}{2}\|\mathbf{M}_i^t \cdot \mathbf{x}_k^t - \mathbf{D}^t\|^2 + \lambda\|\mathbf{X}_k^t\|_1 < \varepsilon$
8: Output:$\mathbf{x}_*^t \leftarrow \mathbf{x}_k^t$

In the algorithm, s represents the number of satellites in one cluster and K represent the number of cycles. The function of soft-thresholding function [11]:

$$\text{soft}(w_i, \lambda) = \text{sign}(w_i) \max(|w_i| - \lambda, 0) \tag{14}$$

After PG we can get an optimal resource allocation method at time slot t and substituting into formula (5) and (6), the performance of the system can be evaluated.

4   Simulation

In our simulations, there is a mega-constellation of LEO satellites deployed at 1000km consist of 300 LEO satellites, spreading across 15 orbits and 20 satellites each orbit. The details of other simulation parameters and their values are shown in Table 2. We Consider a satellites cluster with 5 satellites and the intensity of ground requests is randomly distributed. The channel between the satellite and the user is modeled as the additive white Gaussian noise (AWGN) model.

Table 2. Simulation parameters

Parameters	Values
Number of LEO satellites	300
Number of orbital planes	15
Altitude of the satellite	1000 km
Ka-band frequency f_c	20 GHz
Inclination of orbit	90°
Total satellite power	30 dBw
Half-open angle of each satellite $\theta/2$	45°
Half-open angle of each beam $\gamma/2$	2°
Number of beams each satellite	8
Number of virtual beams each satellite	37
Time-segment	60 s
Number of satellites cluster	5

The performance compared with fixed beam and separative beam hopping method and all simulations are performed on the same conditions.

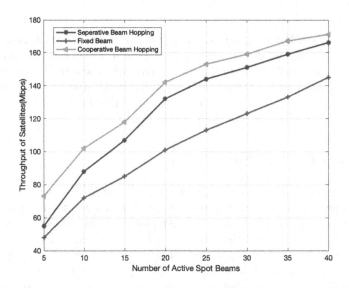

Fig. 4. Throughput of satellite cluster with different number of active beams.

Figure 4 shows the throughput of the system with different beam allocation methods. As the number of active beams increases, proposed method always has better performance than other systems while more beams may reduce the sparsity of system and the effectiveness of the proposed algorithm. Therefore, when all beams are active with insufficient overlapping, cooperative beam hooping only increases limited throughput.

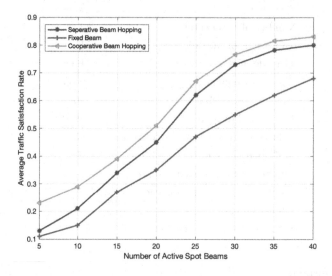

Fig. 5. Average traffic satisfaction rate of the cluster with different number of active beams.

Figure 5 shows the comparison on average traffic satisfaction rate after 10 repeating which has the similar tendency with Fig. 4. Cooperative beam hopping system can dynamically allocate the resource to each beam, but constraint to the number of beams and limited resource it cannot reach perfect satisfaction rate. In conclusion, cooperative beam hopping improves the throughput of system and average traffic satisfaction rate from users.

5   Conclusion and Future Work

In this paper we analyze the cooperative beam hopping for LEO constellation scenario, where the onboard resource is limited, and satellites have much overlapping areas. We model it as a LASSO problem and choose throughput of system and average traffic satisfaction rate of users as indicators to evaluate the performance. After simulation, we find that proposed method has better result on the above measures. However, we ignore the interference between satellites which should be our future focus and the matching theory between terminals and satellites is another question.

Acknowledgements. This study was funded by National Natural Science Foundation of China (grant number 62341105).

References

1. Anzalchi, J., et al.: Beam hopping in multi-beam broadband satellite systems: system simulation and performance comparison with non-hopped systems. In: 2010 5th Advanced Satellite Multimedia Systems Conference and the 11th Signal Processing for Space Communications Workshop, pp. 248–255 (2010). https://doi.org/10.1109/ASMS-SPSC.2010.5586860
2. Hejia, L., Xueliang, S., Ying, C.: 6g very low earth orbit satellite network. HuaweiTech (2022)
3. Hou, K., Yang, J., Liu, F., Zhang, C.: An active queue management algorithm to guarantee the QoS of LEO satellite network. In: 2023 3rd International Symposium on Computer Technology and Information Science (ISCTIS), pp. 1024–1031 (2023). https://doi.org/10.1109/ISCTIS58954.2023.10213116
4. Ivanov, A., Stoliarenko, M., Kruglik, S., Novichkov, S., Savinov, A.: Dynamic resource allocation in LEO satellite. In: 2019 15th International Wireless Communications and Mobile Computing Conference (IWCMC), pp. 930–935 (2019). https://doi.org/10.1109/IWCMC.2019.8766756
5. Li, W., Zeng, M., Wang, X., Fei, Z.: Dynamic beam hopping of double LEO multi-beam satellite based on determinant point process. In: 2022 14th International Conference on Wireless Communications and Signal Processing (WCSP), pp. 713–718. IEEE. https://doi.org/10.1109/WCSP55476.2022.10039244. https://ieeexplore.ieee.org/document/10039244/
6. Lin, Z.: Multi-satellite beam hopping based on load balancing and interference avoidance for NGSO satellite communication systems. IEEE Trans. Commun. **71**(1), 282–295 (2023)
7. Liu, S., Lin, J., Xu, L., Gao, X., Liu, L., Jiang, L.: A dynamic beam shut off algorithm for LEO multibeam satellite constellation network. IEEE Wirel. Commun. Lett. **9**(10), 1730–1733 (2020). https://doi.org/10.1109/LWC.2020.3002846
8. Liu, W., Tian, F., Jiang, Z., Li, G., Jiang, Q.: Beam-hopping based resource allocation algorithm in LEO satellite network. In: Yu, Q. (ed.) SINC 2018. CCIS, vol. 972, pp. 113–123. Springer, Singapore (2019). https://doi.org/10.1007/978-981-13-5937-8_13
9. Sun, Z., Li, T.: Development prospect of mega low earth orbit constellation satellite communication networks. ZTE Technol. J. (2021)
10. Tian, F., Huang, L., Liang, G., Jiang, X., Sun, S., Ma, J.: An efficient resource allocation mechanism for beam-hopping based LEO satellite communication system. In: 2019 IEEE International Symposium on Broadband Multimedia Systems and Broadcasting (BMSB), pp. 1–5 (2019). https://doi.org/10.1109/BMSB47279.2019.8971890
11. Wright, J., Ma, Y.: High-Dimensional Data Analysis with Low-Dimensional Models: Principles, Computation, and Applications. Cambridge University Press, Cambridge (2022)

IP Tunneling Based for Mobility Management in LEO Constellation Networks

Wenqian Wang[1,2], Feng Tian[1,2], Chaoyu Ren[1,3], Zhaolong Ding[1,3], and Zijian Yang[1(✉)]

[1] Innovation Academy for Microsatellites of CAS, Shanghai 201203, China
yanghhh@mail.ustc.edu.cn
[2] University of Chinese Academy of Sciences, Beijing 100049, China
[3] Shanghaitech University, Shanghai 201210, China

Abstract. LEO constellation networks, with their advantages of high throughput, wide coverage and strong robustness, have become a key solution to meet future demands for massive data communications and global coverage. However, due to the high-speed movement of low orbit satellites relative to the ground, mobility management is one of the most challenging research topics for enabling mobility service in LEO satellite networks. In this paper, we propose an IP tunneling based scheme for handling the relative mobility between space and ground. The LEO satellite network consists of the access network and the transmission network. The access network includes the user terminals, the space base station on satellite, and the core network in gateway. The transmission network includes the space router, inter-satellite link, feeder link, and the ground router. In the IP tunneling scheme, the access network and the transmission network belong to two networks respectively, where general routing algorithm is used to handle the relative mobility between the satellites and ground stations, and the IP tunneling is used to handle the relative mobility between the satellites and user terminals. Compared with the Mobile IPv6 (MIPv6), the IP tunneling scheme achieves better performance in terms of signaling overhead, memory overhead and switch delay. Analysis result shows that the IP tunneling scheme can greatly reduce the demand of on-board resources.

Keywords: Mobility Management · IP Tunneling · LEO Satellite Network

1 Introduction

The LEO constellation network has become a promising approach for global communication services, due to its wide coverage and rapid construction [1]. Such as, SpaceX company has lunched more than 5000 satellites [https://www. spacex.com/] and the OneWeb company has lunched more than 450 satellites

Q. Yu (Ed.): SINC 2023, CCIS 2057, pp. 60–67, 2024.
https://doi.org/10.1007/978-981-97-1568-8_6

[https://oneweb.net/]. How to leverage the LEO constellation network for global communication has attracted interests from both academy and industry.

Due to the high-speed relative mobility between the satellite and user terminal, and the limited single LEO satellite coverage, the mobility management is an important issue in LEO constellation network. There have been some researches for this issue. As shown in Table 1, mobility management schemes fall into two categories, centralized and distributed.

Table 1. Categories of mobility management schemes.

Category	Mobility Management technology	Representative
Centralized	Host-based	MIPv6, FMIPv6
	Network-based	PMIPv6, FPMIPv6
Distributed	DMM-based	DIPS

In Centralized Mobility Management (CMM), the mobility information is kept at a single mobility anchor. Data packets are transmitted via this anchor. There are some centralized schemes such as MIPv6 [8] and Fast Mobile IPv6(FMIPv6) [9], etc., which are the host-based mobility management protocols. Furthermore, the network-based mobility management protocols, for instance, PMIPv6 [10], and Fast Proxy Mobile IPv6 (FPMIPv6) [3], have been developed later. The host-based and network-based mobility management protocol are analyzed and compared in terms of handover latency, handover blocking probability, and packet loss in refs. [7]. The result shows that the network-based mobility management outperforms the host-based for the reason that the network-based mobility service provided by Local Mobility Anchor (LMA) and Mobility Access Gate (MAG) reduce the control signals and handover procedures. However, there are some disadvantages of CMM such as non-optimal routes, lack of scalability, single point of failure and duplicate multicast traffic [2].

To solve the problems of CMM, a Distributed Mobility Management (DMM) based scheme in LEO satellites network is proposed in refs. [6]. It deploys the distributed mobility anchors to achieve more optimal routing path, better scalability and robustness. Nevertheless, signaling interactions are frequent in DMM scheme, thus it increases the signaling overhead and complexity of the communication system.

In this paper, we propose an IP tunneling based scheme for handling the relative mobility between space and ground to reduce the demand of on-board resources.

The contributes of this paper are summarized as follows:

1. We propose an IP Tunneling Based for Mobility Management (IPTBMM) in LEO constellation network firstly. For a clear description, we design a LEO satellite communication system shown in Fig. 1. There are two IP address

layer in the proposed LEO satellite communication system. The communication between terminals and data network adopts IP tunnel.

2. We propose for the first time a home satellite maintenance and updating mechanism through cooperation between satellites and ground to reduce the demand of on-board resources.

3. We compare the IP tunneling scheme with the MIPv6 in terms of signaling overhead, memory overhead and switch delay.

The rest of the paper is organized as follows. Section 2 is an overview of IPTBMM. Section 3 analyzes the satellite overhead of IPTBMM and compares it with MIPv6 in terms of resource cost. Finally, the conclusions are given in Sect. 4.

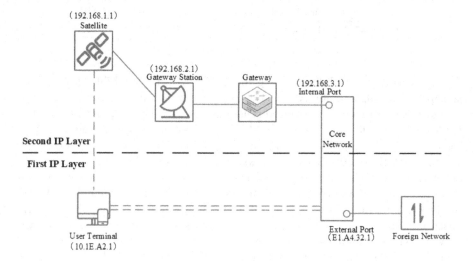

Fig. 1. The proposed LEO communication system.

2   An Overview of IPTBMM

In IPTBMM scheme, the terminal and the external port of the core network belong to the first layer of IP address, and they have length of 8 bytes, where the first 4 bytes are bound to their home core network. Satellites, gateway stations, and internal ports of the core network have the second layer of IP address, and the length is 4 bytes.

The two layers of IP address are independent of each other. As shown in Fig. 2, the data transmission between the terminal and the foreign network is tunnel transmission, in which the satellite and the ground gateway station constitute the transmission layer, and the data transmission process is transparent to the terminal and foreign network.

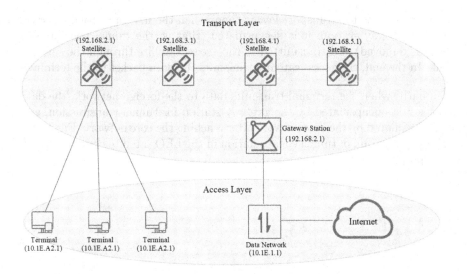

Fig. 2. IP address allocation.

Terminal IP	Home Satellite IP	Refresh Time
A1.B2.10.1	196.168.1.2	T_0
A1.1H.01.1	196.168.2.1	T_0
2B.2C.10.2	196.168.1.3	T_0
......	......	......

Fig. 3. Home satellite table.

The step of IPTBMM is as follows:

1. Switch. After the terminal switches from the first access satellite to the second access satellite, the gateway station notifies the core network of the switching result.
2. Update Routing Information. The core network maintains the terminal location information, and updates the home satellite table according to the handover result. As shown in Fig. 3, the home satellite table includes the IP address of the terminal and the IP address of the home satellite to which the terminal belongs.
3. Tunnel Communication. When the foreign network needs to transmit data to the satellite terminal, after the data arrives at the core network, the core network obtains the IP address of the satellite that the destination terminal belongs to at the current moment by searching the home satellite table. Then the core network encapsulates the data packet (that is, adds the tunnel transmission head) for tunnel transmission. The new IP head is the same IP

address layer as the transport layer. Then, when the new data packet arrives at the gateway station, it is decapsulated (that is, the tunnel transmission head is removed) and transmitted to the access satellite through the satellite link. In the end, the access satellite transmits the data packet to the terminal.

Similarly, when the terminal transmits data to the foreign network, the data packet is first encapsulated by the gateway station for tunnel transmission, and then decapsulated by the core network after reaching the core network. Figure 4 is a schematic diagram of tunnel encapsulation in the LEO satellite communication system proposed in this paper.

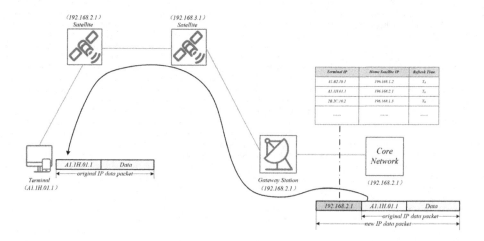

Fig. 4. Tunnel encapsulation in the proposed LEO satellite communication.

3   Cost Analysis

This section will compare IPTBMM and mobility management scheme adopting single IP address layer, for instance, MIPv6 in terms of resource cost.

3.1   Signaling Overhead

Mobility management protocols require various management signaling to provide continuous communication services for terminals that constantly change access points. Signaling overhead is defined as: the sum of control data packets exchanged between the mobile node and the home agent.

The signaling overhead signaling of MIPv6 consists of location registration overhead C^r and location query overhead C_q [5].

In satellite networks, the transmission overhead of datagrams is directly related to the distance from the source node to the destination node, and is proportional to the number of sending location registrations. For the convenience of calculation, it is assumed that the registration and registration confirmation frame sizes are the same as L_r. Define the single bit transmission overhead of control signaling between node A and node B as $C_{A,B}$, the number of location registrations per unit time is k^r, and the gateway station is represented by G. In IPTBMM scheme, the location registration overhead is same as MIPv6. When Mobility Node (MN) corresponds to only one Correspondent Node (CN), the location registration overhead is:

$$C^r_{IPTBMM} = C^r_{MIPv6} = \sum_{i=0}^{k^r} 2 * (C_{MN,G} + C_{MN.CN}) * L_r \qquad (1)$$

In MIPv6 scheme, when CN initiates communication with MN for the first time, there is no user terminal table entry in the sending terminal table. Therefore, a location query request needs to be sent to the gateway station; After the communication between CN and MN is established, MN will register the location with CN. At this time, there is no need to initiate a location query request to the gateway station. In IPTBMM scheme, there is no location query overhead. CN directly sends the data packet to the gateway station, which sends the data to MN.

In a single communication between the MN and CN, the cost of location query is:

$$C^q_{MIPv6} = 2 * C_{MN,G} * L_r \qquad (2)$$

3.2   Memory Cost

In MIPv6 scheme, the gateway station updates the routing table, and broadcasts the updated routing table to all satellites. That means every satellite must store the routing table of the whole network.

In IPTBMM scheme, the maintenance and update of the terminal IP address and the terminal's home satellite are delegated to the core network, and each satellite only needs to maintain the routing information of the next-hop satellite. Memory usage is reduced by $1/3$.

3.3   Switching Delay

In MIPv6 scheme, the handover process can be divided into four stages: link layer handover, mobility detection, duplicate address detection and location update. Figure 5 describes the MIPv6 protocol switching processing diagram [4].

Fig. 5. Tunnel encapsulation in the proposed LEO satellite communication.

The switching delay of MIPv6 is:

$$D_{MIPv6} = D_{L2}^{MIPv6} + D_{MD}^{MIPv6} + D_{DAD}^{MIPv6} + D_{Update}^{MIPv6} \quad (3)$$

Among them, D_{L2} is the link layer switching delay, D_{MD} is the movement detection process delay, D_{DAD} is the duplicate address detection process delay, and D_{Update} is the location update process delay. In single IP address for mobility management, after the gateway station updates the routing table, it will broadcast the updated routing table to all satellites. Thus, D_{Update} of MIPv6 includes the confirmation delay and the distribution delay as formula (4) shown.

$$D_{Update}^{MIPv6} = D_{confirm} + D_{distribution} \quad (4)$$

It is assumed that the distance between adjacent satellites is about 4000 km, the delay of every inter-satellite link transmission is about 12ms. However, in IPTBMM, there is no need to initiate routing table broadcast, only the core network needs to maintain and update routing information. Thus, the location update process delay of IPTBMM include only the confirm delay, which is shown in formula (5). $D(S(BU))$ is the delay of one hop of single binding update packet, $D(S(BAck))$ is the delay of one hop of single binding acknowledge packet. $H(MN, HA)$ is the numbers of hop between MN and Home Agent (HA). In this way, transmission delay and the demand for on-board resources will be reduced.

$$D_{Update}^{IPTBMM} = D_{confirm} = (D(S(BU)) + D(S(BAck))) * H(MN, HA) \quad (5)$$

4   Conclusion

In this paper, we propose an IP tunneling based scheme for mobility management in LEO constellation network. In this scheme, the access network and the transmission network belong to two networks respectively, and the IP tunneling is used to transport data packets. Analysis in Sect. 3 shows that IPTBMM can efficiently reduce or even relieve the work of maintaining terminal routing information of satellites, reduce memory cost and switching delay compared with MIPv6.

Acknowledgments. This paper is supported by National Natural Science Foundation of China (grant number 62341105).

References

1. Al-Hraishawi, H., Chougrani, H., Kisseleff, S., Lagunas, E., Chatzinotas, S.: A survey on non-geostationary satellite systems: the communication perspective. IEEE Commun. Surv. Tutor. **25**, 101–132 (2022)
2. Chan, H., Liu, D., Seite, P., Yokota, H., Korhonen, J.: Requirements for distributed mobility management. Technical report (2014)

3. Chung, J.M., Lee, D., Song, W.J., Choi, S., Lim, C., Yeoum, T.: Enhancements to FPMIPv6 for improved seamless vertical handover between LTE and heterogeneous access networks. IEEE Wirel. Commun. **20**(3), 112–119 (2013)
4. Ding, Y.: Low latency technology for mobility management in LEO satellite. Master's thesis, Xidian University (2020)
5. Guo, Y.: Research on mobility support for LEO satellite networks. Master's thesis, Xidian University (2021)
6. Han, W., Wang, B., Feng, Z., Zhao, B., Yu, W.: Distributed mobility management in IP/LEO satellite networks. In: 2016 3rd International Conference on Systems and Informatics (ICSAI), pp. 691–695. IEEE (2016)
7. He, D., You, P., Yong, S.: Comparative handover performance analysis of MIPv6 and PMIPv6 in LEO satellite networks. In: 2016 Sixth International Conference on Instrumentation & Measurement, Computer, Communication and Control (IMCCC), pp. 93–98. IEEE (2016)
8. Johnson, D., Perkins, C., Arkko, J.: Mobility support in IPv6. Technical report (2004)
9. Koodli, R.: Fast handovers for mobile IPv6. Technical report (2005)
10. Shen, X., Lin, X., Zhang, K.: Encyclopedia of Wireless Networks. Springer, Cham (2020). https://doi.org/10.1007/978-3-319-78262-1

Remote Sensing Image Fusion Method Based on Retinex Model and Hybrid Attention Mechanism

Yongxu Ye⑩, Tingting Wang⑩, Faming Fang$^{(\boxtimes)}$ ⑩, and Guixu Zhang⑩

School of Computer Science and Technology, East China Normal University,
Shanghai 200062, China

51215901060@stu.ecnu.edu.cn, {tingtingwang,fmfang,
gxzhang}@cs.ecnu.edu.cn

Abstract. Pansharpening is a technique that fuses a low-resolution multispectral image (LRMS) and a panchromatic image (PAN) to obtain a high-resolution multispectral image (HRMS). Based on the observation that PAN and LRMS respectively have the characteristics of illumination component and reflection component of HRMS after Retinex decomposition, this paper proposes an inverse Retinex model guided pansharpening network, termed as AIRNet. Specifically, a Spatial Attention based Illuminance Module (SAIM) is proposed to convert the PAN to the illuminance component of HRMS. And a Hybrid Attention-based Reflectance Module (HARM) is used to convert the LRMS to the reflection component of the HRMS. Finally, based on the inverse Retinex model, the corresponding illuminance component and reflection component of the obtained HRMS are fused to obtain HRMS. Qualitative and quantitative comparison experiments with state-of-the-art pansharpening methods on multiple remote sensing image datasets show that AIRNet has significantly outstanding performance. In addition, multiple ablation experiments also show that the proposed SAIM and HARM are effective modules of AIRNet for pansharpening.

Keywords: Pansharpening · Remote sensing image fusion · Spatial attention mechanism · Channel attention mechanism · Inverse Retinex model

1 Introduction

Remote sensing images play an important role in fields such as surface object classification [7] and geographical data [9]. In the remote sensing image imaging process, panchromatic images (PAN) are mixed images obtained by the satellite sensor responding to the light of the entire panchromatic band. It has high spatial resolution, but only has spectral information of a single band. Multi-spectral images (LRMS) are multi-band images obtained by satellite sensors responding to light in multiple different bands. Therefore, they have spectral information in multiple bands, but the spatial resolution is low. In order to obtain high resolution multi-spectral images (HRMS), researchers have

Q. Yu (Ed.): SINC 2023, CCIS 2057, pp. 68–82, 2024.
https://doi.org/10.1007/978-981-97-1568-8_7

proposed a technology that combines LRMS and PAN, called Pansharpening [5]. Pan-sharpening methods can be mainly divided into component substitution (CS) methods, multi-resolution analysis (MRA) methods, variational optimization (VO) methods, and deep learning (DL) methods [23].

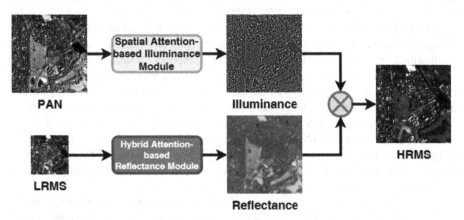

Fig. 1. Example of the Pansharpening method proposed in this paper.

The core idea of the CS methods is to first transform the LRMS to obtain the spatial and spectral components, then replace the spatial components with PAN, and finally inversely transform to obtain the target HRMS. Representative methods mainly include: principal component analysis (PCA) [21] and adaptive Gram-Schmidt (GSA) [2], et al. Generally, the CS methods have less computational complexity, but can easily cause serious loss of spectral information. The main idea of the MRA methods is to decompose the LRMS at multiple resolutions, replace its spatial components with PAN, and finally obtain the target image after inverse decomposition. Classic methods include: additive wavelet brightness proportion method (AWLP) [19], smooth filter intensity modulation method (SFIM) [20] and "à trous" wavelet transform method (ATWT) [24], et al. Compared with CS methods, MRA methods have better spectral preservation capabilities, but often lead to significant spatial distortion. The VO methods build a variational model by assuming a potential relationship between the observation image and the target image, and use the prior information of the image to constrain the model. The optimal solution of the model corresponds to the Pansharpening result. Specific methods can be divided into Bayesian methods [27] and variational methods [10]. The VO methods can maintain the spatial information and spectral information of the HRMS image in a balanced manner, but how to establish the potential relationship between the observation image and the target image and select appropriate prior terms is a big challenge.

In recent years, convolutional neural networks (CNN) have been widely used in different computer vision tasks, and they also shine in Pansharpening tasks [12]. [18] first proposed using CNN for Pansharpening. After that, Hu et al. [11] proposed a multi-scale multi-depth convolutional neural network (MSDCNN) to extract richer features at different scales for Pansharpening. Xu et al. [31] established an efficient observation model for the Pansharpening task and expanded it into a deep network for solution, namely

GPPNN. In addition, Cai and Huang [6] regarded Pansharpening as a task based on super-resolution guidance and proposed a progressive pansharpening network SRCNN. In order to make full use of the shallow features in multi-scale information, Wang et al. [28] used a U-shaped network for Pansharpening. In addition, in order to improve the generalization ability of the model, several methods apply unsupervised strategies such as generative adversarial models [17] for Pansharpening.

In addition, a more interesting idea is to introduce the Retinex model into Pansharpening [8]. The Retinex model decomposes the image into a reflection component and an illumination component [15]. The reflection component reflects the reflectivity of the object to different light, and the illumination component is usually considered to be the result of the interaction between the ambient light source and the geometric structure. Generally, the Retinex model-based Pansharpening method first performs Retinex decomposition on the LRMS to obtain its illumination component, then uses the illumination component and PAN to perform weighted fusion to obtain a new illumination component, and then adds it to the original LRMS to obtain the HRMS. In essence, this type of methods are still CS methods. Therefore, the loss of spectral information is inevitable.

Inspired by the Retinex model-based Pansharpening method, an Attention-based Inverse Retinex Network (AIRNet) guided by Retinex model is proposed in this paper. As shown in Fig. 1, AIRNet inputs PAN and LRMS into Spatial Attention-based Illuminance Module (SAIM) and Hybrid Attention based Reflectance Module (HARM) to obtain the corresponding illuminance and reflectance components of HRMS, and then HRMS is obtained by Retinex decomposition. AIRNet effectively utilizes the rich spatial information in PAN and the rich spectral information in LRMS.

The main contributions of this paper are as follows:

1. The first use of Retinex model for Pansharpening. A novel architecture for Pansharpening is proposed to derive the illuminance and reflection components of HRMS from PAN and LRMS, respectively, and then the HRMS is obtained by Retinex decomposition.
2. A spatial attention-based illuminance module (SAIM) is proposed to transform the PAN with rich spatial information into the corresponding illuminance component of HRMS.
3. A hybrid attention-based reflectance module (HARM) is proposed by combining channel attention and spatial attention, which is used to convert LRMS containing a large amount of spectral information into the corresponding reflectance component of HRMS.
4. Comparison of experimental results with multiple state-of-the-art methods on multiple satellite datasets demonstrates the superiority of the proposed method.

2 Related Works and Motivation

2.1 Attention Mechanism

The human visual perception system does not process the entire scene at the same time, and attention selectively captures salient parts in order to better capture the visual structure [16]. Attention-based networks designed based on this property can utilize useful information in images more effectively.

Spatial Attention Mechanism. As shown in Fig. 2(a), the spatial attention module splices the maximum pooling and average pooling of the input features $\mathbf{F} \in \mathbb{R}^{H \times W \times C_{in}}$ along the channel dimensions and feeds them into the convolutional layer to obtain the intermediate output $\tilde{\mathbf{F}} \in \mathbb{R}^{H \times W \times C_{in}}$, which is then multiplied with the original feature map to obtain the spatial attention map after going through the sigmoid function. The details are as follows:

$$SA(\mathbf{F}) = \mathbf{F} \cdot g(\sigma(\mathbf{W} \otimes [MAX(\mathbf{F}), AVG(\mathbf{F})])) \tag{1}$$

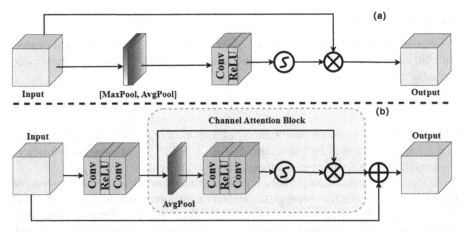

Fig. 2. Overall structure of the different attention mechanisms. (a) spatial attention block; (b) residual channel attention block.

where $g(\cdot)$ and $\sigma(\cdot)$ denote the Sigmoid and ReLU functions, respectively, $\mathbf{W} \in \mathbb{R}^{2 \times k \times k \times C_{out}}$ is a C_{out} convolution kernel of size $k \times k \times 2$, $\otimes$ denotes the convolution operation, and $[\cdot]$, $MAX(\cdot)$, $AVG(\cdot)$ denote the splicing along channels, maximum pooling, and average pooling operations, respectively, and $\cdot$ stands for element-by-element multiplication.

Residual Channel Attention Block. Since the low-resolution space is rich in low-frequency information and some valuable high-frequency information, and each convolution kernel in the convolutional layer has only a local receptive field to the spatial dimension, the use of the channel attention mechanism can take advantage of the interdependence between the feature channels to make the network focus on the more valuable features in the channel dimension. Wang et al. [26] proposed the Residual Channel Attention Block (RCAB), as shown in Fig. 2(b). The residual block and long jump connections enable the backbone network to retain part of the original feature information; the channel attention mechanism can further enhance the fitting ability of the network by extracting the dependencies between channels. Specifically, the structure of RCAB can be represented as:

$$RCAB(\mathbf{F}) = \mathbf{F} + CA(\mathbf{W}^1 \otimes \sigma(\mathbf{W}^2 \otimes \mathbf{F})) \tag{2}$$

where $\mathbf{W}^1 \in \mathbb{R}^{C_{in} \times k \times k \times C_{out}}$, $\mathbf{W}^2 \in \mathbb{R}^{C_{in} \times k \times k \times C_{out}}$ denote two convolutional layers stacked by two convolutional kernels of size $k \times k \times C_{in}$ of C_{out}, respectively. $CA(\cdot)$ is the channel attention module, which has the following structure:

$$CA(\mathbf{F}) = \hat{\mathbf{F}} \cdot g(\mathbf{W}_U \otimes \sigma(\mathbf{W}_D \otimes AVG(\mathbf{F}))) \tag{3}$$

where $\hat{\mathbf{F}}$ denotes the intermediate feature map of the inputs to the channel attention module in RCAB, and $\mathbf{W}_U$ and $\mathbf{W}_D$ denote the convolution kernels used for channel scaling in the channel attention module, respectively.

2.2 Retinex Model

The Retinex model is a color perception model that simulates the human visual system, with the goal of decomposing an observed image $\mathbf{O} \in \mathbb{R}^{H \times W \times C}$ into its illuminance and reflectance components, i.e.:

$$\mathbf{O} = \mathbf{I} \cdot \mathbf{R} \tag{4}$$

where $\mathbf{I} \in \mathbb{R}^{H \times W \times C}$ denotes the illuminance component of the brightness of the object in the scene represented by the image, and $\mathbf{R} \in \mathbb{R}^{H \times W \times C}$ denotes the reflectance component of the reflectivity of the image object to the light.

In studies in the literature [30], the illuminance component is often considered to be the result of the interaction between the ambient light source and the geometric structure, i.e., it possesses a large amount of spatial information. The reflectance component contains more information about the reflectance of an object to different lights, i.e., rich spectral information. Because of this, the Retinex model can be used as an effective feature extractor for both spectral and spatial information. Also Retinex model is widely used in tasks such as image low-light enhancement [22].

2.3 Motivation

In the imaging process of remote sensing images, PAN is a single-band image obtained by absorbing light in panchromatic wavelength bands, which has both light source information and rich geometric structure information; while LRMS is a multispectral image obtained by absorbing light in several different wavelength ranges, which contains rich spectral information, i.e., it reflects the reflectivity of the scene objects to different light. Combined with the fact that the Retinex model decomposes the image into an illuminance component reflecting the interaction between the ambient light source and the geometric structure, and a reflectance component containing the different light reflectance of the object, it is natural to assume that the PAN and the LRMS are the a prior information inherent to the illuminance and reflectance components of the HRMS, respectively. Therefore, HRMS can be obtained by utilizing the powerful fitting ability of neural networks to transform PAN and LRMS into the illuminance and reflectance components of HRMS.

Thanks to the spatial attention and channel attention in the attention mechanism, the above image transformation can be effectively realized. On the one hand, the HRMS illuminance component is obtained from the PAN, which pays more attention to the combination effect of the ambient light source and geometry, so the illuminance module can be constructed based on the spatial attention; on the other hand, in order to obtain the information of the reflectivity of the scene objects to the different wavelengths of the light, it is necessary to deal with the LRMS on a channel-by-channel basis, and for this purpose, a reflectance module based on the mixture of the channel attention and the spatial attention is constructed.

3 Proposed Method

In this section, the proposed AIRNet is described in detail. As shown in Fig. 3, the whole network architecture contains two branches: one branch is used to convert PAN into illuminance component and the other one converts LRMS into reflectance component. Finally, based on the Retinex model, the illuminance component and the reflectance component are multiplied to obtain the HRMS.

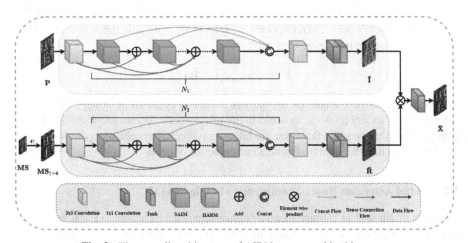

Fig. 3. The overall architecture of AIRNet proposed in this paper.

3.1 Retinex Model Formulation

The HRMS to be solved is denoted as $\mathbf{X} \in \mathbb{R}^{H \times W \times C}$, PAN and LRMS are denoted as $\mathbf{P} \in \mathbb{R}^{H \times W \times 1}$ and $\mathbf{MS} \in \mathbb{R}^{h \times w \times C}$, respectively, and $\mathbf{MS}_{\uparrow \times 4} \in \mathbb{R}^{H \times W \times C}$ denotes the $4\times$ up-sampling of $\mathbf{MS}$. According to the definition of Retinex model, as long as the corresponding illumination component $\bar{\mathbf{I}} \in \mathbb{R}^{H \times W \times C}$ and reflection component $\bar{\mathbf{R}} \in \mathbb{R}^{H \times W \times C}$ are obtained, the target image $\mathbf{X}$ can be obtained. Since the number of channels of $\mathbf{P}$ and $\mathbf{MS}$ are not the same, in order to get the $\bar{\mathbf{I}}$ from $\mathbf{P}$, the number of channels of $\mathbf{P}$ is firstly extended by convolution. i.e.:

$$\hat{\mathbf{P}} = Conv_{\times 1}(\mathbf{P}) \tag{5}$$

where $Conv_{\times 1}(\cdot)$ denotes the convolutional layer, the subscript denotes the number of convolutional layers, and $\hat{\mathbf{P}} \in \mathbb{R}^{H \times W \times C}$ denotes the channel-expanded image. Then $\hat{\mathbf{P}}$ is traversed through N_1 densely connected illuminance modules SAIM, and finally $\bar{\mathbf{I}}$ is obtained by a convolutional layer and a 1×1 convolutional layer with Tanh activation function. The whole process can be represented as:

$$\bar{\mathbf{I}} = Conv_{\times 2}\left(\mathbf{F}_{\times N_1}^{SAIM}\left(\hat{\mathbf{P}}\right)\right) \tag{6}$$

where $\mathbf{F}_{\times N_1}^{SAIM}(\cdot)$ denotes N_1 densely connected illuminance modules.

On the other hand, **MS** is first obtained using bicubic interpolation to obtain $\mathbf{MS}_{\uparrow \times 4}$, and then $\bar{\mathbf{R}}$ is obtained by a similar process as described above.

$$\mathbf{MS}_{\uparrow \times 4} = Bicubic_{\times 4}(\mathbf{MS}) \tag{7}$$

$$\bar{\mathbf{R}} = Conv_{\times 2}\left(\mathbf{F}_{\times N_2}^{HARM}\left(\mathbf{MS}_{\uparrow \times 4}\right)\right) \tag{8}$$

where $Bicubic_{\times 4}(\cdot)$ denotes $4\times$ upsampling using bicubic interpolation, and $\mathbf{F}_{\times N_2}^{HARM}(\cdot)$ denotes N_2 densely connected reflectance modules HARM. Finally the estimated HRMS is obtained based on the Retinex model:

$$\bar{\mathbf{X}} = Conv_{\times 1}\left(\bar{\mathbf{I}} \odot \bar{\mathbf{R}}\right) \tag{9}$$

The entire network architecture employs a densely connection strategy to increase the stability of model training. In addition the intermediate feature maps obtained from multiple illuminance and reflectance modules are subjected to splicing operations along the channel so that the shallow features can be better utilized.

3.2 Spatial Attention Based Illuminance Module (SAIM)

The study by Jin et al. [13] found that the relationship of each channel of $\bar{\mathbf{I}}$ to $\bar{\mathbf{R}}$ should not be portrayed by the same simple function, but by complex functions that are independent of each other. Therefore the designed SAIM treats different channels of $\bar{\mathbf{I}}$ differently while maintaining the interaction between the ambient light source and the geometrical structure. The detailed structure of the SAIM is shown in Fig. 4(a). For the channel-expanded $\hat{\mathbf{P}}$, the channel separation operation is performed first:

$$\{\hat{\mathbf{P}}_1, \cdots, \hat{\mathbf{P}}_C\} = split\left(\hat{\mathbf{P}}^{in}\right) \tag{10}$$

where $\hat{\mathbf{P}}^{in}$ denotes the input feature map. $split(\cdot)$ denotes the channel separation operation. $\{\hat{\mathbf{P}}_1, \cdots, \hat{\mathbf{P}}_C\} \in \mathbb{R}^{H \times W \times 1}$ denotes the single-channel feature map after channel separation. Then the spatial attention mechanism and multilayer convolutional layers are used to realize the interaction between the ambient light source and the geometric structure:

$$SAIM_i\left(\hat{\mathbf{P}}_i\right) = Conv_{\times 3}\left(SA\left(\hat{\mathbf{P}}_i\right)\right) \tag{11}$$

$$\hat{\mathbf{P}}^{out} = Conv_{\times 1}\left(\left[Conv_{\times 1}(SAIM_1), \cdots, Conv_{\times 1}(SAIM_C)\right]\right) \qquad (12)$$

where $SA(\cdot)$ denotes the spatial attention module and $\hat{\mathbf{P}}^{out}$ denotes the output interaction result feature map. This process simulates the interaction between the ambient light source and the geometric structure, and realizes the transformation from $\mathbf{P}$ to $\bar{\mathbf{I}}$. SAIM uses a parameter sharing strategy, which not only reduces the training parameters, but also effectively avoids network overfitting.

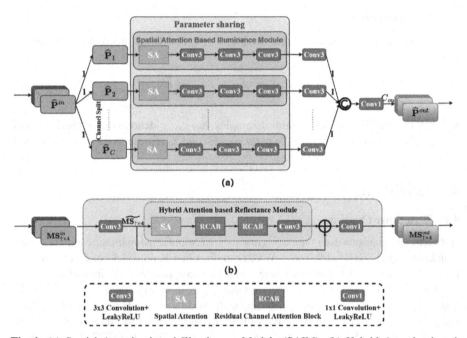

Fig. 4. (a) Spatial Attention-based Illuminance Module (SAIM); (b) Hybrid Attention-based Reflection Module (HARM).

3.3 Hybrid Attention Based Reflectance Module (HARM)

The structure of HARM is shown in Fig. 4(b), which first performs feature pre-extraction operation on the input feature map $\mathbf{MS}^{in}_{\uparrow \times 4}$:

$$\widetilde{\mathbf{MS}}_{\uparrow \times 4} = Conv_{\times 1}\left(\mathbf{MS}^{in}_{\uparrow \times 4}\right) \qquad (13)$$

where $\widetilde{\mathbf{MS}}_{\uparrow \times 4}$ denotes the pre-extracted feature map. Since $\bar{\mathbf{R}}$ contains information about the object reflection of light from different wavelength bands, the SA module is used to focus on a specific object and combined with RCAB to filter the corresponding channels. The work of Woo et al. [29] shows that combining SA and RCAB can yield better results in some tasks. Finally, residual connections are used to retain more input information. The specific process can be represented as:

$$HARM\left(\widetilde{\mathbf{MS}_{\uparrow\times4}}\right) = Conv_{\times1}\left(RCAB_{\times2}\left(SA\left(\widetilde{\mathbf{MS}_{\uparrow\times4}}\right)\right)\right) \tag{14}$$

$$\mathbf{MS}_{\uparrow\times4}^{out} = Conv_{\times1}\left(\widetilde{\mathbf{MS}_{\uparrow\times4}} + HARM\left(\widetilde{\mathbf{MS}_{\uparrow\times4}}\right)\right) \tag{15}$$

where $RCAB_{\times2}(\cdot)$ means execute RCAB twice, and $\mathbf{MS}_{\uparrow\times4}^{out}$ means output feature map.

3.4 Loss Function

The proposed AIRNet network uses the Mean Absolute Error function (MAE) as the loss function, i.e.:

$$L = \frac{1}{N}\left\|\sum_{i=1}^{N}\overline{\mathbf{X}}_{\Theta_{AIRNet(i)}} - \mathbf{X}_i\right\|_1 \tag{16}$$

where N denotes the number of training samples, $\overline{\mathbf{X}}_{\Theta_{AIRNet(i)}}$ denotes the i-th fusion result HRMS obtained by AIRNet after the parameter Θ, and $\mathbf{X}_i$ represents the i-th real labeled data.

4 Experiments

4.1 Experiment Settings

Datasets. The datasets use in the experiments are remote sensing images acquired from QuickBird (QB) and WorldView-2 (WV2) satellites in 4-band and 8-band, respectively, for the MS images. All remote sensing images are cropped into MS images of size 64 × 64 and PAN images of 256 × 256, and then the resolution of the acquired PAN/MS images is reduced by a factor of 4 according to the Wald's protocol [25]. The LRMS/PAN image block size of 16 × 16/64 × 64 for training and 64 × 64 for the real labeled data GT are obtained. The number of image pairs obtained from the QB/WV2 dataset is 10596/9118 pairs, and the dataset is separated at a ratio of 90%/10% to be used for training/validation, respectively.

Evaluation Metrics. In order to validate the performance of the proposed network, a reduced-resolution evaluations in the presence of a reference image and a full-resolution evaluations using real remote sensing data are performed separately. Five widely used quantitative metrics are chosen for the reduced-resolution evaluations, namely SAM [32], SCC [33], QAVE [1], ERGAS [4] and the generic image quality index Q4/Q8 [14]. The QNR [3] metric and its two components D_λ and D_S are then use in the full-resolution evaluations.

Training Details. Experiments are implemented on Nvidia GTX 2080ti GPUs using the Pytorch framework. During training, the epoch is set to 700, the learning rate is 6.5e−4, the learning rate decreases by 0.5 times every 40 epochs, the training batch size is 16, and an Adam optimizer with weight decay, β_1, and β_2 of 9.5e−5, 0.9, and 0.999, respectively, is used.

Table 1. Average results on QuickBird dataset.

	Reduced Resolution					Full Resolution		
	Q4↑	QAVE↑	SAM↓	ERGAS↓	SCC↑	D_λ↓	D_S↓	QNR↑
GSA	0.8723	0.8721	3.0938	2.0194	0.9088	0.0640	0.1027	0.8407
AWLP	0.8665	0.8664	3.1087	2.0155	0.9164	0.0679	0.0948	0.8440
ATWT	0.8611	0.8652	3.2040	2.0378	0.9134	0.0793	0.1106	0.8191
SFIM	0.8542	0.8599	3.2438	2.0966	0.9099	0.0635	0.0870	0.8553
PNN	0.7787	0.7732	3.5218	2.6695	0.8235	0.0523	0.0468	0.9031
MSDCNN	0.8878	0.8895	2.7394	1.9609	0.8989	0.0443	<u>0.0331</u>	0.9244
MUCNN	0.8995	0.8993	2.5666	1.8615	0.9082	0.0506	0.0485	0.9040
SRPPNN	0.9117	0.9137	2.4376	<u>1.7578</u>	0.9131	<u>0.0257</u>	0.0349	<u>0.9407</u>
GPPNN	<u>0.9146</u>	<u>0.9142</u>	<u>2.3593</u>	1.7897	<u>0.9167</u>	0.0546	0.0449	0.9030
Ours	**0.9297**	**0.9301**	**2.1758**	**1.6218**	**0.9287**	**0.0255**	**0.0318**	**0.9427**

Tips: 1. (**Bold**: Best, <u>Underline</u>: second best)

4.2 Comparison with State-of-the-Art

In this section, the proposed AIRNet is compared with other state-of-the-art pansharpening methods, which include four traditional methods: i.e., GSA [2], ATWT [24], AWLP [19], and SFIM [20], in addition to five deep-learning based methods: the PNN [18], MSDCNN [11], MUCNN [28], SRPPNN [6], and GPPNN [31]. In addition to this, we also performed qualitative comparison experiments on the up-sampled version of LRMS (EXP).

Evaluation on Reduced Resolution. Columns 2–6 of Table 1 and Table 2 show the average quantitative results of each method on the QB and WV2 reduced-resolution datasets, respectively. As can be seen from the data in the tables, the method proposed in this paper is significantly better than the other methods. In addition, the methods based on deep unfolding such as GPPNN and SRPPNN are slightly better than the other methods on the QB dataset. Meanwhile, SRPPNN performs better on the WV2 dataset. Fig. 5 and Fig. 6 show the results of fusing a pair of QB test data by each method and the error map between this fusion result and the real image, respectively, and it can be found that our method performs better in both spatial information recovery and spectral information preservation.

Evaluation on Full Resolution. The last three columns of Table 1 and Table 2 give the average quantitative metric results of the various methods on the QB and WV2 full-resolution datasets. It can be seen that on the QB dataset, the best results are obtained for all three metrics of our method. On the WV2 dataset, the QNR still outperforms the other methods, although the results for the D_λ and D_S metrics are not optimal. Figure 7 gives the results of each method fusing a pair of WV2 real data, and this paper's method still has good visual results.

Table 2. Average results on WorldView-2 dataset.

	Reduced Resolution					Full Resolution		
	Q8↑	QAVE↑	SAM↓	ERGAS↓	SCC↑	D_λ↓	D_S↓	QNR↑
GSA	0.9190	0.9135	6.7579	4.0076	0.8967	0.0524	0.1287	0.8258
AWLP	0.8932	0.8920	6.7229	4.4648	0.8855	0.0611	0.1060	0.8394
ATWT	0.8949	0.8951	6.6278	4.4232	0.8869	0.0694	0.1150	0.8237
SFIM	0.8798	0.8819	6.7435	4.6349	0.8878	0.0510	0.1074	0.8471
PNN	0.9485	0.9483	4.8106	2.9017	0.9368	<u>0.0376</u>	0.1031	0.8633
MSDCNN	0.9529	0.9530	4.6903	2.7662	0.9419	**0.0364**	0.0802	0.8863
MUCNN	0.9517	0.9512	4.6655	2.8446	0.9426	0.0425	0.1033	0.8564
SRPPNN	<u>0.9608</u>	<u>0.9608</u>	<u>4.2076</u>	<u>2.5060</u>	<u>0.9527</u>	0.0559	**0.0577**	<u>0.8867</u>
GPPNN	0.9563	0.9556	4.3993	2.6939	0.9481	0.0429	0.0798	0.8807
Ours	**0.9639**	**0.9641**	**3.9955**	**2.4312**	**0.9579**	0.0410	<u>0.0754</u>	**0.8896**

Tips: 2. (**Bold**: Best, <u>Underline</u>: second best)

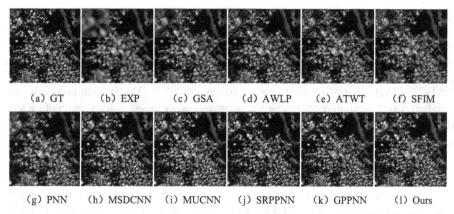

(a) GT   (b) EXP   (c) GSA   (d) AWLP   (e) ATWT   (f) SFIM

(g) PNN   (h) MSDCNN   (i) MUCNN   (j) SRPPNN   (k) GPPNN   (l) Ours

Fig. 5. Qualitative comparison on the reduced-resolution QB dataset. The red box in the upper left corner is a zoomed-in view of the area corresponding to the red box in the image. (Color figure online)

4.3 Ablation Study

The Number of N_1 and N_2. To explore the effect of the number of N_1 and N_2 on network performance, ablation experiments are performed on the WV2 dataset. As shown in Table 3, the network performance is best at $N_1 = 2, N_2 = 2$, and relatively poor at $N_1 = 3, N_2 = 1$ and $N_1 = 3, N_2 = 3$. Therefore, the setting of $N_1 = 2, N_2 = 2$ is used in all the experiments.

The Effectiveness of SAIM and HARM. In order to verify the effectiveness of both SAIM and HARM, the ablation experiments are conducted on the WV2 dataset as well.

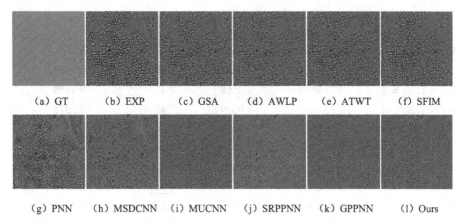

(a) GT   (b) EXP   (c) GSA   (d) AWLP   (e) ATWT   (f) SFIM

(g) PNN   (h) MSDCNN   (i) MUCNN   (j) SRPPNN   (k) GPPNN   (l) Ours

Fig. 6. Error maps corresponding to the results in Fig. 5. For better visualization, all error maps are shown rescaled by adding 0.5 to each pixel.

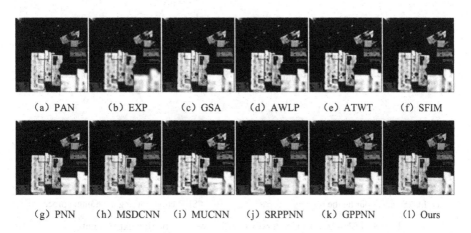

(a) PAN   (b) EXP   (c) GSA   (d) AWLP   (e) ATWT   (f) SFIM

(g) PNN   (h) MSDCNN   (i) MUCNN   (j) SRPPNN   (k) GPPNN   (l) Ours

Fig. 7. Qualitative comparison on the full-resolution WV2 dataset. The red box in the lower right corner is a zoomed-in view of the area corresponding to the red box in the image.

The ablation experiments consider four scenarios, I. Both HARM and SAIM are used; II. SAIM is used without HARM; III. HARM is used but not SAIM and IV. Neither HARM nor SAIM is used. As shown in Fig. 8, I. performed the best. In contrast, the PSNR and LOSS curves of II. and III. performed poorly during training. Meanwhile, III. shows better performance than II., that is, SAIM has less effect on the model than HARM. IV. shows the worst performance, indicating that the proposed SAIM and HARM are effective modules.

Table 3. The average results of ablation experiments on WorldView-2 dataset.

Configurations	PSNR↑	SAM↓	ERGAS↓	SCC↑
$N_1 = 1, N_2 = 1$	33.0422	4.0392	2.4627	0.9569
$N_1 = 1, N_2 = 2$	32.9909	4.0558	2.4721	0.9559
$N_1 = 1, N_2 = 3$	33.0880	4.0199	2.4524	0.9571
$N_1 = 2, N_2 = 1$	33.0359	4.0467	2.4692	0.9567
$N_1 = 2, N_2 = 2$	**33.1551**	**3.9955**	**2.4312**	**0.9579**
$N_1 = 2, N_2 = 3$	28.6520	6.4600	3.9761	0.9212
$N_1 = 3, N_2 = 1$	33.1070	<u>4.0117</u>	<u>2.4425</u>	0.9574
$N_1 = 3, N_2 = 2$	28.7342	6.5128	3.9330	0.9231
$N_1 = 3, N_2 = 3$	<u>33.1078</u>	4.0291	2.4469	<u>0.9577</u>

Tips: 3. (**Bold**: Best, <u>Underline</u>: second best)

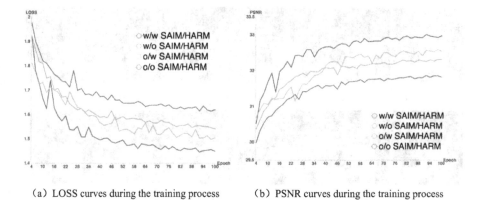

(a) LOSS curves during the training process   (b) PSNR curves during the training process

Fig. 8. PSNR and LOSS curves in effectiveness ablation experiments

5 Conclusion

In this paper, an attention-based pansharpening method, AIRNet, is designed under the guidance of the Retinex model. Specifically, AIRNet uses the Spatial Attention-based Illuminance Module (SAIM) to convert the PAN into the corresponding illuminance component of the HRMS, and uses the Hybrid Attention-based Reflection Module (HARM) to convert the LRMS into the corresponding reflection component of the HRMS, and finally multiplies the illuminance component and the reflection component according to the Retinex model to obtain the HRMS to be solved. Finally, the illuminance and reflectance components are multiplied according to the Retinex model to obtain the HRMS to be solved. Comparison experimental results with other state-of-the-art methods show that the proposed method is more advantageous in qualitative and quantitative comparisons. Future work will focus on how to use the Retinex model to establish the relationship between the observed image and the target image and to propose a more explanatory variational model. Meanwhile, the use of unsupervised strategies will also be explored to increase the generalization ability of the panchromatic sharpening model, and the framework will be applied to a wider range of remote sensing image fusion applications to improve the quality of remote sensing images from different sources, and to contribute to subsequent tasks, such as feature identification, vegetation monitoring, and so on. Tasks such as feature identification and vegetation monitoring.

References

1. Aiazzi, B., Alparone, L., Baronti, S., et al.: MTF-tailored multiscale fusion of high-resolution MS and Pan imagery. Photogramm. Eng. Remote. Sens. **72**(5), 591–596 (2006)
2. Aiazzi, B., Baronti, S., Selva, M.: Improving component substitution pansharpening through multivariate regression of MS + Pan data. IEEE Trans. Geosci. Remote Sens. **45**(10), 3230–3239 (2007)
3. Alparone, L., Aiazzi, B., Baronti, S., et al.: Multispectral and panchromatic data fusion assessment without reference. Photogramm. Eng. Remote. Sens. **74**(2), 193–200 (2008)
4. Alparone, L., Wald, L., Chanussot, J., et al.: Comparison of pansharpening algorithms: outcome of the 2006 GRS-S data-fusion contest. IEEE Trans. Geosci. Remote Sens. **45**(10), 3012–3021 (2007)
5. Xiao, L., Liu, P.F., Li, H.: Progress and challenges in the fusion of multisource spatial-spectral remote sensing images. J. Image Graph. **25**(5), 851–863 (2020)
6. Cai, J., Huang, B.: Super-resolution-guided progressive pansharpening based on a deep convolutional neural network. IEEE Trans. Geosci. Remote Sens. **59**(6), 5206–5220 (2020)
7. Cao, X., Yao, J., Xu, Z., et al.: Hyperspectral image classification with convolutional neural network and active learning. IEEE Trans. Geosci. Remote Sens. **58**(7), 4604–4616 (2020)
8. Dong, W., Xiao, S., Li, Y., et al.: Hyperspectral pansharpening based on intrinsic image decomposition and weighted least squares filter. Remote Sens. **10**(3), 445 (2018)
9. Du, P., Liu, S., Xia, J., et al.: Information fusion techniques for change detection from multitemporal remote sensing images. Inf. Fusion **14**(1), 19–27 (2013)
10. Fang, F., Li, F., Shen, C., et al.: A variational approach for pan-sharpening. IEEE Trans. Image Process. **22**(7), 2822–2834 (2013)
11. Hu, J., Hu, P., Kang, X., et al.: Pan-sharpening via multiscale dynamic convolutional neural network. IEEE Trans. Geosci. Remote Sens. **59**(3), 2231–2244 (2020)
12. Yong, Y.: Deep-learning approaches for pixel-level pansharpening. Natl. Remote Sens. Bull. **26**(12), 2411–2432 (2023)
13. Jin, X., Gu, Y., Liu, T.: Intrinsic image recovery from remote sensing hyperspectral images. IEEE Trans. Geosci. Remote Sens. **57**(1), 224–238 (2018)
14. Krizhevsky, A., Sutskever, I., Hinton, G.E.: ImageNet classification with deep convolutional neural networks. In: Advances in Neural Information Processing Systems, vol. 25 (2012)
15. Land, E.H., McCann, J.J.: Lightness and retinex theory. Josa **61**(1), 1–11 (1971)
16. Larochelle, H., Hinton, G.E.: Learning to combine foveal glimpses with a third-order Boltzmann machine. In: Advances in Neural Information Processing Systems, vol. 23 (2010)
17. Ma, J., Yu, W., Chen, C., et al.: Pan-GAN: an unsupervised pan-sharpening method for remote sensing image fusion. Inf. Fusion **62**, 110–120 (2020)
18. Masi, G., Cozzolino, D., Verdoliva, L., et al.: Pansharpening by convolutional neural networks. Remote Sens. **8**(7), 594 (2016)
19. Otazu, X., González-Audícana, M., Fors, O., et al.: Introduction of sensor spectral response into image fusion methods. Application to wavelet-based methods. IEEE Trans. Geosci. Remote Sens. **43**(10), 2376–2385 (2005)

20. Schowengerdt, R.A.: Remote Sensing: Models and Methods for Image Processing. Elsevier, Amsterdam (2006)
21. Shah, V.P., Younan, N.H., King, R.L.: An efficient pan-sharpening method via a combined adaptive PCA approach and contourlets. IEEE Trans. Geosci. Remote Sens. **46**(5), 1323–1335 (2008)
22. Song, X., Huang, J., Cao, J., et al.: Multi-scale joint network based on Retinex theory for low-light enhancement. Signal Image Video Process. **15**, 1–8 (2021)
23. Vivone, G., Alparone, L., Chanussot, J., et al.: A critical comparison among pansharpening algorithms. IEEE Trans. Geosci. Remote Sens. **53**(5), 2565–2586 (2014)
24. Vivone, G., Restaino, R., Dalla Mura, M., et al.: Contrast and error-based fusion schemes for multispectral image pansharpening. IEEE Geosci. Remote Sens. Lett. **11**(5), 930–934 (2013)
25. Wald, L., Ranchin, T., Mangolini, M.: Fusion of satellite images of different spatial resolutions: assessing the quality of resulting images. Photogramm. Eng. Remote. Sens. **63**(6), 691–699 (1997)
26. Wang, F., Jiang, M., Qian, C., et al.: Residual attention network for image classification. In: Proceedings of the IEEE Conference on Computer Vision and Pattern Recognition, pp. 3156–3164 (2017)
27. Wang, T., Fang, F., Li, F., et al.: High-quality Bayesian pansharpening. IEEE Trans. Image Process. **28**(1), 227–239 (2018)
28. Wang, Y., Deng, L.J., Zhang, T.J., et al.: SSconv: explicit spectral-to-spatial convolution for pansharpening. In: Proceedings of the 29th ACM International Conference on Multimedia, pp. 4472–4480 (2021)
29. Woo, S., Park, J., Lee, J.-Y., Kweon, I.S.: Cbam: Convolutional block attention module. In: Ferrari, V., Hebert, M., Sminchisescu, C., Weiss, Y. (eds.) ECCV 2018. LNCS, vol. 11211, pp. 3–19. Springer, Cham (2018). https://doi.org/10.1007/978-3-030-01234-2_1
30. Xu, J., Hou, Y., Ren, D., et al.: Star: a structure and texture aware retinex model. IEEE Trans. Image Process. **29**, 5022–5037 (2020)
31. Xu, S., Zhang, J., Zhao, Z., et al.: Deep gradient projection networks for pan-sharpening. In: Proceedings of the IEEE/CVF Conference on Computer Vision and Pattern Recognition, pp. 1366–1375 (2021)
32. Yuhas, R.H., Goetz, A.F.H., Boardman, J.W.: Discrimination among semi-arid landscape endmembers using the spectral angle mapper (SAM) algorithm. In: JPL, Summaries of the Third Annual JPL Airborne Geoscience Workshop. Volume 1: AVIRIS Workshop (1992)
33. Zhou, J., Civco, D.L., Silander, J.A.: A wavelet transform method to merge Landsat TM and SPOT panchromatic data. Int. J. Remote Sens. **19**(4), 743–757 (1998)

A Multi-branch Hierarchical Feature Extraction Network Combining Sentinel-1 and Sentinel-2 for Yellow River Delta Wetlands Classification

Xinhao Li, Mingwei Liu, Qingwen Dou, Mingming Xu[✉], Shanwei Liu, and Hui Sheng

Land Surveying and Mapping Institute of Shandong Province, Jinan 250102, China
xumingming@upc.edu.cn

Abstract. The classification of wetlands in the Yellow River Delta is important for the monitoring of vegetation dynamics, rational resource utilization, and ecosystem protection. In this paper, the multi-temporal Sentinel-1 and Sentinel-2 data from 2021 are used to extract 168 features about spectral, index, texture, and polarization scattering. And based on the multi-source features, a novel multi-branch hierarchical feature extraction network (MHFE) is designed to classify the wetlands in the Yellow River Delta. By virtue of the multi-branch characteristics, the proposed MHFE can target the processing data with different features. The network includes the attention convolution module and fuzzy information module designed according to the characteristics of the data. The results show that the overall accuracy of multi-source features can reach 87.55% when classifying collaboratively, which is significantly higher than that of single-source features, and the fusion of multi-source data helps to improve the accuracy of wetland classification. Comparing with a variety of advanced deep learning classifiers, the proposed MHFE has the highest overall accuracy, which verifies that the application of this model to classify the wetlands in the Yellow River Delta has validity.

Keywords: Multi-source data fusion · Self-attention mechanism · Wetland classification · Yellow River estuary

1 Introduction

Wetlands, along with oceans and forests, comprise the three major global ecosystems. Wetlands are renowned as the kidneys of the Earth, natural reservoirs, and biodiversity hotspots. Coastal wetlands, located in the transitional zones between land and sea, demonstrate distinctiveness in terms of maintaining ecological balance, providing resource support, improving environmental quality, reducing disaster risks, and promoting socio-economic development. The Yellow River Delta wetland is one of the world's largest, most well-preserved, and youngest temperate coastal wetlands. However, under the dual influence of human activities and climate change, the wetlands of the Yellow River Delta are facing increasingly serious threats, including shrinking in size, systematic degradation, and deteriorating habitats. Therefore, conducting detailed classification and accurate monitoring of the wetlands in the Yellow River Delta is an essential prerequisite for subsequent management and protection efforts.

© The Author(s), under exclusive license to Springer Nature Singapore Pte Ltd. 2024
Q. Yu (Ed.): SINC 2023, CCIS 2057, pp. 83–99, 2024.
https://doi.org/10.1007/978-981-97-1568-8_8

As a result of the launching and networking of Earth observation satellites, multi-source remote sensing data have been gradually applied to the classification of coastal wetlands. Optical imagery provides rich spectral information but is limited by cloudy, rainy, and foggy conditions, resulting in reduced availability and making it challenging to distinguish land cover with similar spectral reflectance. Synthetic Aperture Radar (SAR) has strong penetration capabilities and is less affected by weather conditions. It can reflect vegetation structural features, detect hydrological features, and also overcome the limitations of optical images in monitoring coastal wetlands. Single-temporal remote sensing data are not seasonal and cyclical, supplemented with multi-time-phase remote sensing data can reflect the climatic and temporal nature. Studies have shown that integrating multi-temporal images results in higher overall classification accuracy compared to single-temporal images [1]. This indicates that the collaborative classification of multi-temporal optical data and radar data is a favorable choice for enhancing the accuracy of wetland land cover classification [2].

Before the rise of deep learning technology, wetland classification was often constrained by methods that combined feature information with machine learning classification. Wetland classification in the Yellow River Delta region was achieved through the collaborative use of radar data and multispectral data using methods such as Maximum Likelihood, Decision Trees, and Support Vector Machine [3]. Zhang [4] used a random forest algorithm combined with object-oriented to realize the accurate classification of typical wetland vegetation in the Yellow River Mouth Protected Area. By integrating multispectral remote sensing data, SAR data, and topographic data, wetlands were accurately extracted using the Random Forest algorithm, and the classification map can be directly employed for sustainable management, ecological conservation, and evaluation of coastal wetlands [5].

With the development of relevant technologies, deep learning has gained widespread attention and is being applied in various fields. Therefore, constructing deep classification architectures to achieve fine-grained classification of coastal wetlands has distinct advantages and development potential. To fully exploit the multi-source data features, researchers often extract the potential information of the image using preprocessing, such as vegetation index, red edge index, texture features, polarization features. Based on the five types of feature indicators, such as water body index extracted from Sentinel-1 and Sentinel-2 time-series data, the Random Forest algorithm, Support Vector Machine algorithm, and Deep Neural Network algorithm were used for comparative classification, and the purpose of the exploration was to find the optimal combination of features and classification strategies for wetland vegetation classification [6]. Han [7] conducted feature exploration using a Convolutional Neural Network (CNN) based on feature intersection learning. The research indicates that this model exhibits good effectiveness and generalization in the scenario of coastal wetland classification with high spectral and multispectral fusion.

However, existing deep learning networks often do not do feature differentiation for the input feature data, and extract features according to the unified features directly, and lack of network design for feature extraction of different data. Designing a network suitable for multi-source data containing various features is essential for wetland classification. Therefore, we have proposed a multi-branch hierarchical feature extraction network (MHFE)designed for different feature data. The different branches of the network can handle input data with different features, enabling targeted and efficient feature extraction. At the same time, we have designed two modules based on the characteristics of the data. These are the attention convolution module (ACM), which increases the focus on high-weight pixels, and the fuzzy information module (FIM), which reduces the impact of noisy data. The main contributions can be summarized as follows:

1) MHFE has the ability to branch to process multiple feature data through multiple branches. It can selectively handle data with a large amount of information and rich details, data with low information content and simplicity, data with prominent features, and data with high noise levels and unclear information.
2) The ACM possesses efficient capability for extracting detailed features. It can enhance the attention to important information and is suitable for processing feature data rich in details.
3) The FIM can extract global information from feature data and effectively reduce the introduction of noise. It can be used for processing feature data with blurred details and high levels of noise.

The remaining sections of this paper are organized as follows: In Section 2, the extent of wetlands in the Yellow River Delta and the use of multi-source data are expounded. Section 3 describes the details of proposed method MHFE. The experimental evaluation are presented in Sect. 4. Finally, Sect. 5 provides a conclusion to this paper.

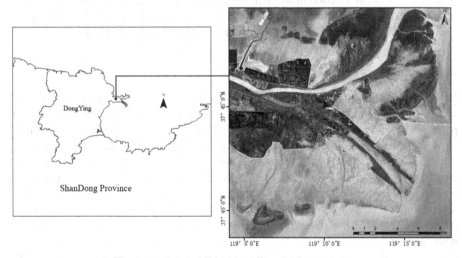

Fig. 1. The geographical location of the study area

2 Study Area and Data

2.1 Study Area

This study selected the entrance of the Yellow River Delta into the sea (119°4′E-119°18′E, 37°38′N-37°49′N) as the research area. It is located between Bohai Bay and Laizhou Bay in a mid-latitude region with a temperate continental monsoon climate characterized by cold winters and hot summers, with rainfall occurring during the warm season. The annual average precipitation in this area is 560 mm. The common wetland vegetation types include native vegetation such as Phragmites australis, Tamarix chinensis, Suaeda and willow forests, as well as the introduced exotic species of Spartina alterniflora. Figure 1 shows the geographical location of the study area.

2.2 Data

This study utilizes two main categories of data: satellite data and sample point data. The satellite data includes Sentinel-1 and Sentinel-2 data, all of which were downloaded from the Copernicus Open Access Hub. Sentinel-1 consists of two satellites, S1A and S1B, which orbit at an altitude of approximately 700 km. They have a spatial resolution of 5 m × 20 m and revisit the same area every 6 days. Sentinel-1 is equipped with a SAR that provides C-band radar data in dual polarization. It is not restricted by cloud cover, rainfall, or lighting conditions, enabling it to provide images regardless of the time of day or weather conditions. Sentinel-2 carries the Multispectral Imager (MSI), which provides high-resolution multispectral imagery up to 290 km wide, covering 13 spectral bands. Sentinel 2 parameters for each band are shown in Table 1.

2.3 Data Preprocessing

Get 2021 Sentinel-1 IW Level-1 Ground Range Detected (GRD) data for 4 views in spring, summer, autumn and winter. Extraction of backward scattering coefficients and polarisation features from Sentinel-1 data, with pre-processing operation steps including Orbit Correction, Thermal Noise Removal, Radiometric Calibration, Refined Lee Filtering, Terrain Correction, Polarisation Decomposition, and Cropping. To acquire cloud-free or low-cloud Sentinel-2 Multi Spectral Instrument (MSI) Level-2A data for the year 2021, covering all four seasons, and perform preprocessing operations including Resampling, Band Synthesis, and Cropping. Project the spectral data and SAR data into the WGS 1984 UTM Zone 50N coordinate system. Finally, the spectral data were aligned with the SAR data pixel by pixel to obtain a multi-source feature dataset. Sample point data, combined with on-site survey photos, high-resolution images from Google Earth, and historical records, were manually interpreted. Pure image elements were selected to classify eight wetland feature types, namely, Suaeda salsa, Natural willow forests, Miscanthus, Reeds, Tamarisks, Arable land, Water bodies, Tidal flats, with a total number of 16,102 samples. Table 2 provides detailed information about the research data, while Table 3 presents details about the training, validation, and testing samples.

Table 1. Sentinel-2 Spectral Band Information

Waveband Number	Sentinel-2A		Sentinel-2B		Spatial Resolution (m)	Main Application
	Centre Wavelength (nm)	Bandwidth (nm)	Centre Wavelength (nm)	Bandwidth (nm)		
1	442.7	21	442.7	21	60	Aerosol
2	492.4	66	492.1	66	10	Blue
3	559.8	36	559.0	36	10	Green
4	664.6	31	664.9	31	10	Red
5	704.1	15	703.8	16	20	Red Edge 1
6	740.5	15	739.1	15	20	Red Edge 2
7	782.8	20	779.7	20	20	Red Edge 3
8	832.8	106	832.9	106	10	NIR
8A	864.7	21	864.0	22	20	Narrow NIR
9	945.1	20	943.2	21	60	Water vapour
10	1373.5	31	1376.9	30	60	SWIR–Cirrus
11	1613.7	91	1610.4	94	20	SWIR 1
12	2202.4	175	2185.7	185	20	SWIR 2

Table 2. Detailed information on the study data

Seasonality	Sentine-1	Sentine-2
Winter	2021-02-20	2021-02-24
Spring	2021-05-06	2021-05-06
Summer	2021-07-25	2021-07-23
Autumn	2021-10-13	2021-10-15

3 The Proposed Method

3.1 Feature Extraction

Spectral bands provide detailed surface information. Vegetation index, water body index can effectively differentiate between different types of land cover. Differences in vegetation growth in response to the red-edge index. Texture features describe the properties of the spatial distribution and arrangement of various details and structures in an image. Radar images are not easily disturbed by the external environment and are sensitive to information on vegetation structure. Polarisation features improve accuracy of land cover classification. Coastal wetlands are mainly composed of vegetation and water. Therefore, in order to characterise the wetland vegetation, hydrology and soil, this study uses

Table 3. The detailed information of training, testing and validation sample

Ground Truth	No.	Color	Class	Train.	Val.	Test.
Yellow River Delta Dataset	1		Water body	355	355	6395
	2		Spartina alterniflora	128	128	2300
	3		Phragmites australis	68	68	1224
	4		Tamarix chinensis	5	6	97
	5		Suaeda salsa	35	35	632
	6		Tidal flat	187	187	3370
	7		Natural willow forest	20	19	346
	8		Cultivated land	7	7	128
			Total	805	805	14492

preprocessed four Sentinel-1 images and four Sentinel-2 images to extract the spectral bands, water index, vegetation index, red-edge index, texture features and polarisation features for each image. In this case, except for the radar backscatter coefficient and polarisation features extracted from the Sentinel-1 data, the rest of the features are extracted or calculated from the Sentinel-2 data. Specific information for each categorical feature is shown in Table 4.

Extracted ten spectral bands of Sentinel-2, with the removal of the coastal/aerosol B1 band, the water vapour B9 band and the cirrus B10 band. Seven vegetation, water, and soil indices reflecting the growth status, health, and climatic cycle of green vegetation, water distribution, and soil coverage were extracted. Extraction of 10 red-edge indices expressing plant leaf area and chlorophyll sensitivity. Extracted 8 texture features that present inter-pixel relationships and patterns. In order to avoid data redundancy, after principal component analysis of the spectral bands, the first principal component with the most information content was taken to do the grey scale covariance matrix GLCM processing. Seven polarisation features mapping the structure and spatial distribution of plant communities were extracted. There are a total of 168 features. The combination of multi-source remote sensing data and multi-feature information data complements each other, enhances the distinguishability of features, and has the prospect of stable reliability and wide application.

3.2 Introduction to the MHFE Network

The multi-feature combination of feature data in this paper consists of multiple sources of satellite imagery, texture, polarization, and other multiple-feature information superimposed on the same spatial location. However, the design of existing deep learning classification networks tends to focus on single-feature data, which is challenging to fit the multi-feature data requirements and cannot effectively perform feature extraction. To better carry out feature extraction, we designed an MHFE that can be more targeted to multi-feature data for feature extraction based on the data characteristics, which contains an ACM and a FIM designed based on the data characteristics.

3.2.1 Introduction of the General Network

The MHFE network structure is shown in Fig. 3. The input data can be roughly categorized into four types of feature data superimposed as shown in Fig. 2, which are $A \in R^{H \times W \times A1}$, $B \in R^{H \times W \times A2}$, $C \in R^{H \times W \times A3}$, $D \in R^{H \times W \times A4}$, where H is the image length and W is the image width; $A1, A2, A3, A4$ is the number of bands of the four types of data. First, we preprocess the data, respectively, the four types of data in accordance with the proportion of their bands, the dimensionality reduction. The feature maps after dimensionality reduction are A', B', C', D', where the bands of A', B', C', D' are $B1$, $B2, B3, B4$; $B1: B2: B3: B4 = A1: A2: A3: A4$.

Table 4. Introduction of Characteristic Variables

Feature index	Index abbreviation	Description of features	Number of features
Spectral feature group	Band	B2, B3, B4, B5, B6, B7, B8, B8a, B11, B12	40
Water body-vegetation-soil index group	NDVI	(B8 − B4)/(B8 + B4)	28
	DVI	B8 − B4	
	RVI	B8/B4	
	EVI	2.5(B8 − B4)/(B8 + 6.0B4−7.5B2 + 1)	
	SAVI	(B8 − B)/(B8 + B4 + 0.5) × (1 + 0.5)	
	NDWI	(B3 − B8)/(B3 + B8)	
	MNDWI	(B3 − B11)/(B3 + B11)	
Red border index group	NDVI_re1	(ρREG4 − ρREG1)/(ρREG4 + ρREG1)	40
	NDVI_re2	(ρREG4 − ρREG2)/(ρREG4 + ρREG2)	
	NDVI_re3	(ρREG4 − ρREG3)/(ρREG4 + ρREG3)	
	ND_re1	(ρREG2 − ρREG1)/(ρREG2 + ρREG1)	
	ND_re2	(ρREG3 − ρREG1)/(ρREG3 + ρREG1)	

(continued)

Table 4. (*continued*)

Feature index	Index abbreviation	Description of features	Number of features
	CI_re	B8/ρREG3−1	
	CI_Green	B8/B3−1	
	RNDVI	(ρREG1 − B4)/(ρREG1 + B4)	
	RRI_1	B8/ρREG1	
	RRI_2	ρREG1/B4	
Texture feature group	GLCM_M	Mean	32
	GLCM_V	Variance	
	GLCM_H	Homogeneity	
	GLCM_C	Contrast	
	GLCM_D	Dissimilarity	
	GLCM_E	Entropy	
	GLCM_SM	Second Momen	
	GLCM_C	Correlation	
Radar polarisation feature group	S1_VV	VV	28
	S1_VH	VH	
	S1_VV-VH	VV-VH	
	S1_VV/VH	VV/VH	
	S1_A	Alpha	
	S1_α	Anisotropy	
	S1_H	Entropy	

Subsequently, a total of four branches were designed to handle different multi-featured data better. The input data for branch 1 is A^', respectively, passed through the convolution module and the ACM, and the output feature matrices are operated on each other. Branch 1 is designed mainly for feature maps with more details, such as Fig. 2(a), where the main information extraction is carried out by convolution and ACM, focusing on the ACM to highlight the important information, which ensures that more information is extracted and reduces the missing information. The input data for branch 2 is B', which is subjected to the convolution module, multi-head self-attention module, and batch normalization, respectively. Branch 2 mainly targets the data with more prominent features, as in Fig. 2(b).The input data of branch 3 are C^', respectively, through the spatial attention module, the multi-head self-attention module, and the convolution module. Branch 3 is mainly for the image data that is more single prominent data, such as texture information, as in Fig. 2(c). The input data of branch 4 are D^', respectively, through the ACM, FIM. Branch 4 mainly focuses on feature data with fuzzy data, as in Fig. 2(d), and extracts only its global features, while reducing the noise in

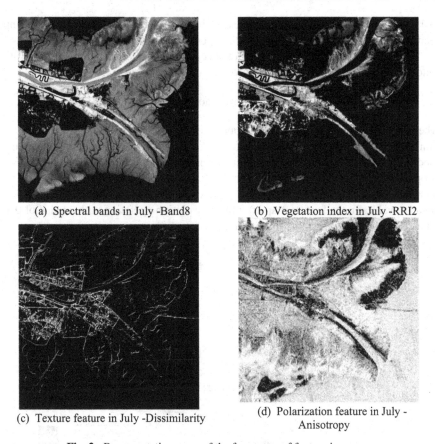

(a) Spectral bands in July -Band8

(b) Vegetation index in July -RRI2

(c) Texture feature in July -Dissimilarity

(d) Polarization feature in July - Anisotropy

Fig. 2. Representative maps of the four types of feature images

the extracted data. The input data of branch 5 is the non-preprocessed data X, which passes through the convolution module, the multi-head self-attention module, and the fully-connected layer. Branch 5 makes up for the missing global information due to the classification process and the lost information such as spatial due to dimensionality reduction by processing the unprocessed image data.

The feature maps obtained through each of the five branches are $A' \in R^{H \times W \times C}$, $B' \in R^{H \times W \times C}$, $C' \in R^{H \times W \times 1}$, $D' \in R^{H \times W \times 1}$, $X' \in R^{H \times W \times C}$. Subsequently, the A', B' is spliced in the spectral dimension and combined with C', D', X', respectively. The network structure design is shown in Fig. 3.

3.2.2 Attention Convolution Module

For informative images, single or simple convolutional kernel superposition is not good enough to extract effective information, so designing the attention convolutional module to ensure the efficient extraction of information. As shown in Fig. 4, the ACM is composed of three branches, a, b, and c, respectively, which play an important information extraction, comprehensive information extraction, play the role of residuals, set the input feature map represented by X, the specific operation is as follows:

The branch a consists of a convolutional kernel, a batch normalization, an activation function Relu, and a fully connected layer as follows:

$$X_1 = FC(Relu(Batch\_norm(Conv2D(X)))) \tag{1}$$

where X_1 is the feature matrix of the output of branch 1; Conv2D($\cdot$)is a 2D convolution operation; Batch_norm($\cdot$) is the batch normalization; Relu($\cdot$) is an abbreviation for modified linear unit, an activation function commonly used in neural networks; FC($\cdot$) is operated through the full connectivity layer. The purpose of the branch design is to focus on the important information by using the activation function Relu to mask the negative values in the feature map after the convolution operation with a value of 0 and retain only the non-zero value information.

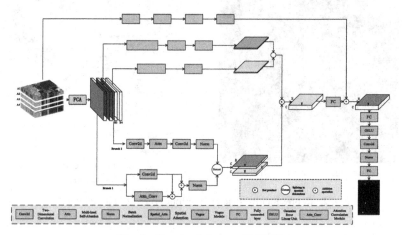

Fig. 3. Structural design of MHFE

The branch b consists of a convolutional, a batch normalization, and a fully connected layer, and the branch c plays the role of the residuals, with a moderating role using the FC linearization layer. Finally, the feature maps of the three branches are summed up and the network structure is shown in Fig. 3. The specific operations are as follows:

$$X_2 = FC(Conv2D(Batch\_norm(X))) \tag{2}$$

$$X_3 = FC(X) \tag{3}$$

$$X = X_1 + X_2 + X_3 \tag{4}$$

where X_2, X_3 are the feature matrices of the outputs of branches b, c.

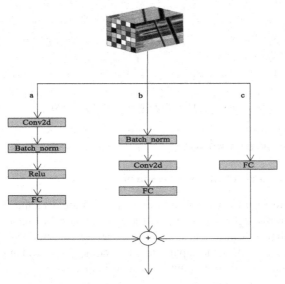

Fig. 4. Attention convolution module design

3.2.3 Fuzzy Information Module

In processing images with little information and a lot of noise in the data, too much noise is often introduced if the information is extracted directly. Therefore, the fuzzy information module is designed with the aim of extracting the main information of the data to minimize the effect of noise as much as possible. As shown in Fig. 5, the core idea of this module is to reduce the spatial size of the feature map patch while extracting the features by convolution, and at the same time, bilinear interpolation is carried out, which is used to recover the spatial size of the feature map patch, so as to realize the role of expanding the main features of the patch and reducing the influence of the noise, which is operated as follows:

$$X' = \text{GELU}(\text{Conv2d}(X)) \tag{5}$$

$$X = \text{Interpolate}(X') \tag{6}$$

$$X = \text{Interpolate}(\text{Conv2d}(X)) \tag{7}$$

where GELU($\cdot$) is a Gaussian error linear unit, which is often used as a nonlinear transformation function in deep learning, and Interpolate ($\cdot$) is a bilinear interpolation operation; X is a feature patch with a spatial size of 7×7, and is a patch that has been convolved with a spatial size of 5×5.

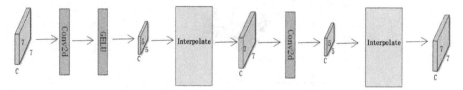

Fig. 5. Fuzzy information module design

4   Experimental Evaluation

4.1   Classification and Accuracy Assessment

In order to explore the effectiveness of multi-source feature data, the classical machine learning random forest algorithm was used to design three comparison schemes, as shown in Table 5. Sentinel-1 feature data, Sentinel-2 feature data, Sentinel-1 and Sentinel-2 feature data were used as the input dataset to classify the wetland features in the Yellow River estuary using the random forest algorithm, respectively. Confusion matrix is usually calculated to visualize and evaluate the performance of the classification model quantitatively, and in this part, the confusion matrix is used to calculate four indexes, namely, Overall Accuracy (OA), Kappa coefficient, Producer Accuracy (PA), and User Accuracy (UA), to evaluate the accuracy of the classification results of the experimental scheme. The Sentinel-1 features are backscattering coefficients and polarization features, and the Sentinel-2 features are raw spectral bands, vegetation index, water index, soil index, red edge index, and texture features.

Table 5. Design of data set input scheme

Input dataset	Characteristics
Features data for Sentinel-1	S1(backscattering coefficient, polarization characteristics)
Features data for Sentinel-2	S2(original spectral bands, vegetation index, water index, soil index, red edge index, texture characteristics)
Features data for Sentinel-1 and Sentinel-2	S1_S2(backscattering coefficient, polarization characteristics, original spectral bands, vegetation index, water index, soil index, red edge index, texture characteristics)

The overall accuracies of the three different input datasets are shown in Table 6. The overall classification accuracies of the multi-source data over the single-source data increased by 11.82% and 2.22%, respectively, and the Kappa coefficients increased by 16.26% and 2.82%, respectively. Tamarix chinensis is the feature with the lowest precision in the single Sentinel-1 SAR data. The low classification accuracy of a feature also lowers the overall accuracy. The large field area of spartina alterniflora can reach

96.79% in PA and 89.40% in UA because spartina alterniflora usually grows in water and has abundant branches and leaves, and its branching structure and wet environment cause the scattering characteristics of radar waves to be different from those of the surrounding water or other features, thus showing distinctive features in the radar data; the classification accuracy of the four types of features, namely, natural willow forest, tamarix chinensis, cultivated land, and tidal flats, in the single Sentinel-2 optical data is greatly improved compared with that of the single SAR data, and the PA can reach up to 88.23% on average, and the UA can reach up to 75.31% on average. The reason is that high-resolution optical data are rich in spectral and textural features, which make it relatively easy to identify and distinguish feature types with different color tones, structural morphology, and coverage forms; in Sentinel-1 and Sentinel-2 multi-source feature data, there are considerable PAs and UAs for each feature type, which neutralizes the inadequacy of a single source of data for classifying a feature.

Overall, multi-source feature data classification is greatly improved compared with single optical data and single SAR data classification. Optical remote sensing data provide rich visual information of the earth's surface, and when optical remote sensing data are limited, radar data can be used as an effective supplementary data source, and radar features can be integrated into the optical features to improve the classification accuracy. At the same time, multi-source data can be combined in many ways and multi-temporal data can be acquired conveniently, so the application of wetland classification in the Yellow River Delta under multi-source feature data is more flexible and more widely applicable. Therefore, based on the multi-source feature remote sensing data, the research on the classification method of wetlands in the Yellow River Delta is carried out.

Table 6. Evaluation of classification accuracy of different datasets

Feature category	Sentinel-1		Sentinel-2		Sentinel-1 + Sentinel-2	
	PA%	UA%	PA%	UA%	PA%	UA%
Suaeda salsa	89.21	89.58	95.44	88.80	95.85	92.77
natural willow forest	66.37	70.75	88.50	78.13	89.38	75.37
spartina alterniflora	96.79	89.40	100.00	98.90	99.51	99.63
phragmites australis	73.87	80.99	87.65	94.13	89.31	93.07
tamarix chinensis	13.89	45.45	69.44	62.50	63.89	71.88
cultivated land	72.22	35.14	100.00	100.00	100.00	100.00
tidal flat	69.50	51.01	94.98	60.06	97.85	65.65
water	71.51	84.99	73.23	97.68	78.37	98.87
OA%	75.73%		85.33%		87.55%	
Kappa coefficient%	66.97%		80.41%		83.23%	

4.2 Comparison of Different Methods

To evaluate the effectiveness and superiority of MHFE, several advanced and widely used deep learning methods are used for comparison based on multi-source feature datasets. The methods include multi-scale 3-D deep convolutional neural network (M3D-DCNN) [8], a CNN-based 3-D deep learning approach(3D-CNN) [9], an attention-based bidirectional long short-term memory network (AB-LSTM) [10], a deep feature fusion network (DFFN) [11], a transformer-based backbone network named SpectralFormer (SPEFORMER) [12], SSFTT [13] and Group-Aware Hierarchical Transformer (GAHT) [14].

To ensure fair performance across models, all deep learning classification methods were run within the Pytorch framework. The minimum batch size and epoch were set to 300 and 300, respectively, to update all model parameters. The optimizer and learning rate scheduler are the same as the original settings in order to obtain the best performance of the other models reported in their papers. For MHFE, we used a stochastic gradient descent (SGD) optimizer with a momentum of 0.9 and a weight decay of 0.0001 to update the training parameters and set the learning rate to a constant of 0.001. The experiments were accelerated using NVIDIA GeForce RTX 3060 GPUs equipped with 12 GB of memory. The classification performance of each model on the Yellow River Delta dataset was evaluated using three metrics: overall accuracy (OA), average accuracy (AA) and Kappa coefficient (Kappa). The classification results of different methods are shown in Table 7.

Table 7. Classification results of different classification methods on multi-source feature dataset

Class	M3D-DCNN	3D-CNN	AB-LSTM	DFFN	SPEFORMER	SSFTT	GAHT	Proposed
1	99.06	98.80	98.62	99.31	97.14	99.94	99.52	99.47
2	100.00	99.74	96.91	99.83	100.00	100.00	100.00	100.00
3	95.26	94.36	89.46	88.24	89.13	92.97	95.67	96.25
4	84.54	5.15	9.28	80.41	83.51	83.51	81.44	92.59
5	95.25	95.89	93.35	96.52	97.78	95.25	93.51	96.30
6	96.26	80.86	89.44	97.39	98.28	96.41	96.44	99.97
7	86.71	49.13	88.15	84.39	78.61	89.60	85.55	82.86
8	100.00	100.00	84.38	100.00	100.00	100.00	100.00	97.89
OA (%)	97.69%	92.47%	94.24%	97.41%	96.70%	97.98%	97.84%	98.80%
AA (%)	94.63%	77.99%	81.20%	93.26%	93.06%	94.71%	94.02%	95.67%
K × 100	96.77%	89.35%	92.01%	96.38%	95.41%	97.17%	96.98%	98.33%

The experimental results show that MHFE outperforms other classification methods. 3D-CNN extracts features in three dimensions, which is one more dimension of feature extraction than 2DCNN. M3D-DCNN captures different levels of information through multi-scale modules, and the densely connected layer in the structure promotes the transfer of information and the reuse of features, and the OA of the two classification methods is 92.47% and 97.69%, respectively; AB-LSTM improves the modeling ability of key information by introducing an attention mechanism to enhance the model's attention to different locations of the input data, thus improving the modeling ability of key information, with an OA of 94.24%; DFFN utilizes dense connections in the network to promote information transfer and feature fusion ability, with the OA increasing to 97.41%; SPEFORMER combines the self-attention mechanism of Transformer and the CNN's spatial information processing, with an OA of 96.70%; SSFTT features a self-supervised learning approach, with a performance excellent classification effect, with an OA exceeding SPEFORMER's by 1.28%; GAHT utilizes a generative adversarial network model, with a higher performance and classification advantage, with an OA of up to 97.84%; MHFE designed a multibranch structure, including an ACM and a FIM, with the highest accuracy among several methods, with an OA as high as 98.80%. In AA, MHFE improves 1.04%, 17.68%, 14.47%, 2.41%, 2.61%, 0.96%, and 1.65% compared to M3D-DCNN, 3D-CNN, AB-LSTM, DFFN, SPEFORMER, SSFTT, and GAHT, respectively; and in Kappa coefficients, MHFE compares favorably with M3D-DCNN, 3D-CNN, AB-LSTM, DFFN, SPEFORMER, SSFTT, and GAHT by 1.56%, 8.98%, 6.32%, 1.95%, 2.92%, 1.16%, and 1.35%, respectively, and AA also presents the same enhancement results. It can be seen that the multi-branch feature module obtains enough feature information, and branch-by-branch information extraction facilitates the improvement of classification accuracy of multi-source feature data.

4.3 Ablation Experiments

We performed ablation experiments on our dataset in order to visualize the role of different modules. This section focuses on the following two main modules: the ACM, and the FIM.

Table 8. Effect of different modules on classification accuracy

Vague Module	Attention Convolution Module	AA	OA	Kappa
√	√	95.67%	98.80%	98.33%
√	×	95.20%	98.18%	97.46%
√	×	89.47%	98.28%	97.60%
×	√	93.50%	98.03%	97.24%
×	×	91.24%	97.55%	96.57%

4.3.1 Attentional Convolution Module

Explore the contribution of the ACM to the MHFE. In order to verify the advancement of the attentional convolution module, the attentional convolution module was removed and replaced with the normal convolution module (shown in green) and the removal of this module (shown in red). As shown in Table 8, both types of experiments showed a significant decrease in accuracy after removing the LMHSA. This phenomenon indicates that the ACM possesses more powerful feature extraction ability than the ordinary convolution module, and also proves that the design of the structure of the ACM is reasonable.

4.3.2 Fuzzy Information Module

Explore the contribution of the FIM to MHFE. The FIM is removed from the whole network. As shown in Table 8, there is a significant reduction in the classification accuracy after removing the FIM. The results show that the FIM allows the model to extract the main information and reduce the introduction of noise when confronted with less informative and noisy data. Therefore, the FIM has a positive contribution to MHFE.

5 Conclusion

This study first demonstrates the positive effect of multi-source feature data fusion on improving classification accuracy. Secondly, based on multi-source feature data, a multi-branch hierarchical feature extraction network MHFE is proposed for wetland classification in the Yellow River Delta. The proposed network utilizes multiple branches to mine the feature distribution, which solves the problem of a single network structure being too deep. The network uses an ACM to extract more representational features, which amplifies the details of the features, and a FIM to reduce the introduction of noise and enhance feature discrimination. Extensive experiments on the Yellow River Delta coastal wetland dataset show that the proposed method outperforms several other leading classification methods.

References

1. Xie, S.Y., Fu, L.B., Li, Y., et al.: Classification method on marsh wetlands in Honghe national nature reserve based on multi-dimensional remote sensing images. Wetland Sci. **19**(01), 1–16 (2021)
2. Ming, Y.S., Liu, Q.H., Bai, H., et al.: Classification and change detection of vegetation in the Ruoergai Wetland using optical and SAR remote sensing data. Nat. Remote Sens. Bull. **27**(6), 1414–1425 (2023)
3. Li, P., Li, D.H., Li, Z.H., et al.: Wetland classification through integration of GF-3 SAR and Sentinel-2B multispectral data over the Yellow River Delta. Geomatics Inf. Sci. Wuhan Univ. **44**(11), 1641–1649 (2019)
4. Zhang, C.Y., Chen, S.L., Li, P., et al.: Spatiotemporal dynamic remote sensing monitoring of typical wetland vegetation in the current Huanghe river estuary reserve. Haiyang Xuebao **44**(1), 125–136 (2022)
5. Xing, H., Niu, J., Feng, Y., et al.: A coastal wetlands mapping approach of Yellow River Delta with a hierarchical classification and optimal feature selection framework. Catena **223**, 106897 (2022)

6. Zhang, L., Luo, W.T., Zhang, H.H., et al.: Classification scheme for mapping wetland herbaceous plant communities using time series Sentinel-1 and Sentinel-2 data. Nat. Remote Sens. Bull. **27**(06), 1362–1375 (2023)

7. Han, Z., Gao, Y., Jiang, X., et al.: Multisource remote sensing classification for coastal wetland using feature intersecting learning. IEEE Geosci. Remote Sens. Lett. **19**, 1–5 (2022)

8. He, M., Li, B., Chen, H.: Multi-scale 3D deep convolutional neural network for hyperspectral image classification. In: 2017 IEEE International Conference on Image Processing (ICIP), pp. 3904–3908. IEEE (2017)

9. Hamida, A.B., Benoit, A., Lambert, P., et al.: 3-D deep learning approach for remote sensing image classification. IEEE Trans. Geosci. Remote Sens. **56**(8), 4420–4434 (2018)

10. Mei, S., Li, X., Liu, X., et al.: Hyperspectral image classification using attention-based bidirectional long short-term memory network. IEEE Trans. Geosci. Remote Sens. **60**, 1–12 (2021)

11. Song, W., Li, S., Fang, L., et al.: Hyperspectral image classification with deep feature fusion network. IEEE Trans. Geosci. Remote Sens. **56**(6), 3173–3184 (2018)

12. Hong, D., Han, Z., Yao, J., et al.: SpectralFormer: rethinking hyperspectral image classification with transformers. IEEE Trans. Geosci. Remote Sens. **60**, 1–15 (2021)

13. Sun, L., Zhao, G., Zheng, Y., et al.: Spectral–spatial feature tokenization transformer for hyperspectral image classification. IEEE Trans. Geosci. Remote Sens. **60**, 1–14 (2022)

14. Mei, S., Song, C., Ma, M., et al.: Hyperspectral image classification using group-aware hierarchical transformer. IEEE Trans. Geosci. Remote Sens. **60**, 1–14 (2022)

A Reliability-Amended-Based Controller Placement Method for LEO Satellite Networks

Shuotong Wei[1,2], Tao Dong[1,2(✉)], Hang Di[1,2], Zhihui Liu[1,2], and Shichao Jin[1,2]

[1] State Key Laboratory of Space-Ground Integrated Information Technology, Space Star Technology Co., Ltd., Beijing 100095, China
dongtaoandy@163.com
[2] Beijing Institute of Satellite Information Engineering, Beijing 100095, China

Abstract. LEO satellite networks are becoming research hot and the utilization of software defined network (SDN) technology brings benefits in flexible network control. Controller placement is important for SDN-enabled networks. In this paper, the impact of delay and reliability on controller placement for SDN-enabled LEO satellite networks is analyzed firstly, following which a reliability-amended model is proposed based on programmable switches distributed on some satellites. The latency-reliability utility function and programmable switch-based enhanced controller placement algorithm are presented. Finally, the effectiveness of the proposed method is proved through simulation and the results show that about 10% reliability of controllers can be improved compared to that without amendment.

Keywords: LEO satellite networks · software defined network (SDN) · controller placement · joint delay and reliability optimization · reliability amendment

1 Introduction

Low earth orbit (LEO) satellite networks own relatively low altitudes, guaranteeing low latency compared with geostationary orbit (GEO) satellites. In the last decade, the number of deployed satellites has increased dramatically and the 3rd Generation Partnership Project proposes non-terrestrial network to take advantage of the convergence of satellite networks and terrestrial 5G networks to incorporate a hybrid satellite network to overcome the limitations in terms of smaller coverage of terrestrial cellular communications and greater terrain restrictions, in which LEO satellite networks are indispensable [1]. However, with the increasingly various services in demand, a huge amount of data is generated, which features fast speed, diversity, sophistication, and heterogeneity. Compared with the terrestrial network, LEO satellite networks computing, storage and other resources are limited. The dynamically changing topology causes frequent routing switching, resulting in difficulties in flexible network control and efficient resource utilization.

Software defined network (SDN) technology is an approach to solve above problems, by decoupling the data plane and control plane [2]. SDN enables communication

© The Author(s), under exclusive license to Springer Nature Singapore Pte Ltd. 2024
Q. Yu (Ed.): SINC 2023, CCIS 2057, pp. 100–114, 2024.
https://doi.org/10.1007/978-981-97-1568-8_9

between controllers and switches by OpenFlow protocol. From the perspective of network architecture, SDN is endowed with flexibility, open interfaces, and centralized control, which solves the problem of rigidity of satellite networks, simplifying network management.

Constrained by on-satellite limited resources and LEO satellite network scale, choosing a suitable set of controllers is of great significance for network performance improvement [3]. Many factors, such as the harsh space environment, frequent link switch and unbalancing traffic, affect the reliability of nodes and links, which will lead to network failure. Therefore, the reliability problem for the control plane is prominent. In order to improve the reliability of satellite networks, SDN-based satellite networks are mainly faced with the controller placement problem (CPP), which directly affects the performance of the network, and at the same time, it is crucial to achieve the minimization of the node-to-controller latency. In summary, it is of important significance to determine the impact of latency and reliability on networks by optimizing the location of the controllers in the LEO satellite network topology.

Latency and reliability are important factors affecting controller placement, due to network latency affects the data processing efficiency and real-time performance. Meanwhile, considering the high dynamics of LEO satellites, the stability of LEO satellite networks will be vulnerable to deterioration. However, unlike the terrestrial CPP, the dynamics of satellite networks and the limited processing resources of satellites prevent existing research on terrestrial CPP from being applied to LEO satellite networks.

To tackle the CPP for LEO satellite networks, researchers conduct some studies that mainly focus on minimizing the average network latency, maximizing network reliability, or optimizing load balancing [2–12]. Research [2] presented an entirely new problem of joint placement optimization for satellite gateways and controllers to maximize network reliability. It obtained the maximum average reliability, given the latency constraint. However, it is only focusing on GEO satellite and terrestrial networks. Study [4] focused on minimizing the average failure probability of the control path, ensuring reliability with certain latency constraints. However, it only considered the satellite network with a few LEO satellite nodes. D. K. Luong et al. in [5] accomplished optimal network reliability with latency constraints by state-of-the-art simulated annealing (SASA), while inter-satellite links are neglected. On the aspects of minimizing network latency, Cheng Chi et al. in [3] considered control latency and load balance to evaluate the average network response control latency, without reliability factors. In research [6], it achieved network reliability maximization and proposed simulated annealing partition-based K-Means (SAPKM), while just considering propagation latency in networks. Study [7] comprehensively investigated the influence of overall latency and proposed a clustering-based network partition algorithm (CNPA) to shorten the maximum end-to-end latency, while seldomly considering the reliability. B. Li et al. in [8] proposed based on a multi-agent deep Q-learning networks method to optimize CPP including latency, load balancing, and path reliability. But they only considered the network in terrestrial networks. Research [9] constructed an optimal space control network to improve the temporal effectiveness of network control, just minimizing the number of controllers with a reliability guarantee. J. Guo et al. in [10] only considered the latency's influence. In [11] they proposed a static placement with a dynamic assignment (SPDA) method to reduce switch-controller latency and optimize load balancing performance, leaving reliability out of consideration. In [12], the controller placement with traffic load was considered to ensure the delivery latency, without reliability considered.

Although providing significant insights, most research separated the latency and reliability to evaluate their impact on CPP. We couple the reliability and latency in the SDN-enabled LEO satellite network. This paper maximizes the joint latency-reliability utility function, with the failure probability decay model. The main contributions of this paper are as follows.

Firstly, a reliability-amendment CPP model is provided for LEO satellite networks, in which programmable switches are distributed to conduct network monitoring and load balancing, hence enhancing reliability. Secondly, a joint latency-reliability utility function (LRUF) is introduced to characterize the overall performance of the satellite network in terms of CPP, in which the correlation between latency and reliability is analyzed, formulating the failure probability as a function of latency. Finally, a programmable switch-based enhanced controller placement algorithm (PECPA) is proposed to achieve the maximum of LRUF.

This paper is organized as follows: Section 2 introduces the system model. In Sect. 3, we formulate the problem and propose PECPA. Section 4 describes the simulation and analysis. We conclude the paper and look forward to future work in Sect. 5.

2   System Model

2.1   LEO Satellite Network Scenario

The investigated satellite network scenario is software defined network (SDN)-based LEO satellite networks, in which the functions of network control and network switch are separately forming the control plane and data plane respectively, as shown in Fig. 1. Meanwhile, in the LEO satellite networks, some nodes in the data plane possess programmability are called as programmable switching nodes (PSNs) in this paper.

The satellite network is denoted as an undirected graph $G(V, E)$, where the node set is $V = \{V_s, Con\}$, representing the set of all satellite nodes with the number of satellite nodes N. And the set of links is E, in which the weight of the link presents the distance between nodes.

The satellite switch node set is denoted as V_s with each element v_s and the total number is n_s. The collection of satellite SDN controllers is Con, and the number is m. The satellite nodes with programmability forming the set $V_{INT} = \{v_{INT}\}$ with total number $|V_{INT}| = n_{INT}$.

In this paper, the controller placement problem (CPP) is studied considering the reliability of nodes and links in LEO satellite networks. The failure probability of node s and link l are denoted as P_v and P_l respectively.

The detailed notations and definitions used in this paper are summarized in Table 1.

2.2   Network Model

Latency Among Controllers and Switches. In SDN-enabled LEO satellite networks, for many services with low latency requirements, the control plane requires control data

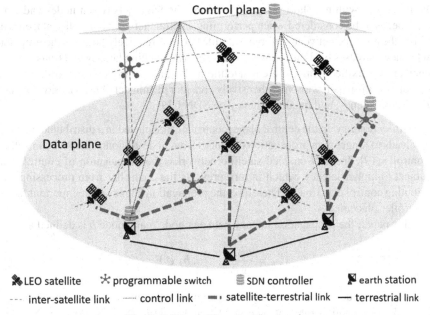

Fig. 1. LEO satellite network scenario.

Table 1. Notations and definitions.

Notation	Definition
$G(\mathbf{V}, \mathbf{E})$	satellite network with node set $\mathbf{V}$ and edge set $\mathbf{E}$
N	number of satellites
$\mathbf{V}_S$	set of satellite switch nodes
$\mathbf{Con}$	set of SDN controllers
$\mathbf{V}_{INT}$	set of PSNs
v_s	satellite switch node
v_{INT}	PSN
m	number of SDN controller
n_s	number of satellite switch nodes
n_{INT}	number of PSNs
P_v	the failure probability of node
P_l	the failure probability of link
nor_R	normalized reliability
nor_L	normalized delay
η	latency reliability utility function

to be processed within a short period. Therefore, the latency between nodes and controllers needs to be considered when performing satellite network controller placement. Besides, the distance between different satellite nodes is relatively large, as the propagation latency increases by 3 ms for every 1000 km. The queuing latency and transmission latency are diversified from different satellite nodes and controllers with traffic loads. Therefore, in this paper, we comprehensively consider the impact of propagation latency, transmission latency, and queuing latency on SDN CPP.

Propagation Latency. In the control plane, controllers are placed in a distributed manner, which leads to longer data propagation distances from node-to-controller and controller-to-controller [9]. In SDN-enabled satellite networks, the propagation of control data is subject to higher latency, which in turn prevents the controller from processing and distributing control data timely. Therefore, the propagation latency is an important factor in controller placement.

In this paper, the propagation latency between node a and node b is defined as

$$\tau_{a,b}^{p} = \frac{dis_{a,b}}{v_{a,b}}, \forall (a, b) \in \mathbf{E}. \tag{1}$$

In (1), $dis_{a,b}$ denotes the propagation distance between node a and node b. $v_{a,b}$ denotes the propagation rate between node a and node b. Besides, $dis_{i,j} = \sum dis_{a,b}$ represents the general propagation distance from a node i to a non-adjacent node j, where the number of inter-satellite links passing through is n_p, and the propagation rate is $v_{i,j} = \sum v_{a,b}/n_p$. The corresponding propagation latency is $\tau_{i,j}^{p} = dis_{i,j}/v_{i,j}$. Furthermore, we define the average propagation latency from node i to other $N - 1$ nodes as

$$\tau_{i}^{p} = \frac{1}{N - 1} \sum \tau_{i,j}^{p}. \tag{2}$$

It represents the property of the node i in propagation latency. Assuming the maximum average propagation latency as $dlp = \max\{\tau_{i}^{p}\}$.

Transmission Latency. In the control plane, controllers are placed in a distributed manner, which leads to Due to the uneven distribution of service flows and data received, some switches have higher loads and need to transmit larger amounts of data to corresponding nodes. In case of state synchronization, more data needs to be transmitted between controllers. When the control plane manages the data plane, there is a high transmission latency in both controllers and nodes. Therefore, the transmission latency is also an important factor that affects the overall performance of the network in CPP.

In LEO satellite networks, the transmission latency from node a to node b is defined as

$$\tau_{a,b}^{t} = \frac{D_{a,b}}{B_{a,b}}, \forall (a, b) \in \mathbf{E}, \tag{3}$$

where $D_{a,b}$ denotes the data size from node a to node b, and $B_{a,b}$ represents the transmission rate from node a to node b, defined as follow

$$B_{a,b} = f\left(B_{\text{band}}, P_{\text{trans}}, N_0, G_{\text{gain}}\right)$$

$$= B_{\text{band}} \log_2\left(1 + P_{\text{trans}}\frac{G_{\text{gain}}g_{a,b}}{dis_{a,b}^2}/(N_0 B_{\text{band}})\right). \tag{4}$$

G_{gain} denotes the interline-of-sight communication coefficient. $g_{a,b}$ means the channel power gain. B_{band} represents the channel bandwidth of inter-satellite link. P_{trans} denotes the transmission power. N_0 is channel power spectral density [13].

Assuming the satellite node i needs to transmit data to n_t nodes, the average node transmission latency as a measure of the average amount of data that the node i needs to transmit during the time period to express the property of the node i in transmission latency, defined as

$$\tau_i^t = \frac{1}{n_t} \sum \tau_{i,j}^t. \tag{5}$$

Besides, set the maximum average transmission latency as $dlt = \max\{\tau_i^t\}$.

Queuing Model. In LEO satellite networks, assuming that there are n nodes and m controllers connected in a satellite network, controller requests are sent randomly from any node, and the probability of a request arriving at the controller is Poisson distributed [14]. Assuming the arrival rate be λ_i, and the total arrival rate of the system is $\lambda = \sum \lambda_i$. Before a request is sent to a processor in a controller, it is stored in the controller's queue. Let the service rate of each controller be μ_i, and the total service rate be $\mu = \sum \mu_i$. Since both service time and arrival time follow an exponential distribution and the number of controllers is m, the number of servers is m, and the system in the SDN-enabled satellite network is a $M/M/m$ queuing model [7]. In this paper, assume that the service rate of all controllers is same, then the service intensity is $\rho = \lambda/m\mu$. The steady-state probability is denoted as

$$p_k = p_0 \left(\frac{\lambda}{\mu}\right)^k \frac{1}{m!m^{k-m}}, k \leq m, \tag{6}$$

while k is the number of services in the queue. As $\sum P_k = 1$, p_0 can be expressed as

$$p_0 = \left[\sum_{k=0}^{m-1} \frac{(m\rho)^k}{k!} + \frac{(m\rho)^m}{m!(1-\rho)}\right]^{-1}. \tag{7}$$

Queuing Latency. In LEO satellite networks, controllers and nodes need to process data streams from multiple nodes simultaneously without instantaneity due to limited capacity. Assuming the stream arrivals in satellite networks obey the Poisson process, the queuing latency of the whole network is the $M/M/m$ model. The queuing latency of the node i is expressed as

$$\tau_i^{\text{que}} = \frac{L_i}{\lambda} = \frac{(m\rho)^m \rho}{m!(1-\rho)^2\lambda}p_0, \tag{8}$$

where L_i denotes the length of queue in node i. More details of the $M/M/m$ model and correlation formula derivation can be referred to [14]. We define the maximum queuing latency is $dlq = \max\{\tau_i^{\text{que}}\}$.

Reliability Among Controllers and Switches. Not only low latency but also high reliability is required when evaluating the performance of satellite networks. The better the reliability of the control plane, the longer the entire network is able to transfer data between nodes without errors. Reliability involves node reliability and link reliability, and controller placement should be placed on nodes with high node reliability and high reliability of links to other nodes [9].

Nodes and Links Reliability. In this paper, the overall reliability of a node is related to the node reliability and the link reliability of other nodes connected to it. Assume the number of inter-satellite links passing between node u to controller node c is n_l, the number of satellite nodes passing through them is n_r, the failure probability of satellite nodes is P_v, the failure probability of inter-satellite links is P_l, then the reliability of links from node u to node c is

$$R_{uc} = \prod_{e \in E_{u \to c}}^{n_l} \left(1 - P_l^e\right) \prod_{v \in V_{u \to c}}^{n_r} \left(1 - P_v^n\right). \tag{9}$$

Furthermore, the number of node u to other nodes is n_u, and the average reliability of node u is

$$R_u = \frac{1}{n_u} \sum R_{uc}. \tag{10}$$

When a controller node is placed on a node with high reliability, it is able to send control flow stably in real time and relatively close to other nodes with low propagation and transmission latency, guaranteeing the real-time performance of control flow. In this paper, in order to better measure the reliability of the satellite node i, the normalized reliability is defined as

$$nor_{Ri} = \frac{R_i}{R_{\max}} \tag{11}$$

where $R_{\max} = \max\{R_i\}$ is the maximum average reliability under a certain time slice of satellite network topology.

Reliability Decay. With each additional hop of the link, the nodes have to reallocate more resource to ensure the same reliability as the previous one. In time slices, the latency grows linearly as the number of nodes on the end-to-end link increases and the end-to-end latency unreliability increases exponentially [15]. It further follows that the reliability of nodes and links decreases exponentially as the latency grows linearly where the probability of not being able to transmit properly is a latency-dependent function. The failure probability of node after decay is expressed as

$$P_v' = f(P_v, \tau) = \left[P_v + \theta_v e^{\tau_v^{\text{que}} + \tau_v^t}\right]_0^1, \tag{12}$$

where θ_v denotes the coefficient of node failure probability with latency, and function $[f]_0^1$ denotes that its value is 1 if $f \geq 1$ and otherwise its value equals to f itself. The failure probability of node increases exponentially as its queuing latency and transmission latency increase. Similarly, the failure probability of link after decay is expressed as

$$P'_l = f(P_l, \tau) = \left[P_l + \theta_l e^{\tau_{a,b}^p + (\tau_a^t + \tau_b^t)/2} \right]_0^1, \tag{13}$$

where θ_l denotes the coefficient of link failure probability with latency. As a link connect node a and node b, the failure probability of a link increases exponentially with propagation latency and transmission latency increase.

3 The Proposed Method

3.1 Problem Formulation

Reliability Amendment. To accurately real-time sense network performance, optimize load balancing, effectively reduce congestion, and further improve the reliability of the entire network, the PSNs are considered to distribute in LEO satellite networks. PSN possesses in-band network telemetry (INT) ability and can achieve network monitoring with low bandwidth consumption. Owing to the programmability, it has prowess in flexibly configuring the network and accurately adjusting traffic load [16]. Therefore, nodes endowed with programmable switch can reduce the failure probability of satellite networks.

In this paper, PSNs are deployed in the satellite network and balance the traffic load to improve the reliability of the nodes and links. There are n_{INT} PSNs, the set of nodes is V_{INT}. Assuming it has the ability to correct the failure probability of nodes and links up to N_{INT} hops, the farther apart, the smaller the amendment of the failure probability. Therefore, the amended failure probability can be expressed as

$$P_{\text{new}} = P_{\text{old}} - \delta(P), \tag{14}$$

where $\delta(P)$ is amendatory factor with $\delta(P) < 1$, $\delta(P)$ is a function of the latency and the number of hops, as shown in (15). Δ is the amendatory increment, θ_2 denotes the decay coefficient with latency. With the certainty of hops, amendatory factor decreases with the increase of latency.

$$\delta(P) = \Delta - (n - N_{\text{INT}})e^{-\tau}\theta_2. \tag{15}$$

Therefore, node-amended failure probability is a function of queuing latency and transmission latency

$$P_{v,\text{new}} = \left[P_{v,\text{old}} - \Delta + (n - N_{\text{INT}})e^{-(\tau_v^{\text{que}} + \tau_v^t)}\theta_2 \right]^+. \tag{16}$$

Similarly, link-amended failure probability is a function of propagation latency and transmission latency

$$P_{e,\text{new}} = \left[P_{e,\text{old}} - \Delta + (n - N_{P4})e^{-(\tau_{a,b}^p + (\tau_a^t + \tau_b^t)/2)}\theta_2 \right]^+. \tag{17}$$

Latency-Reliability Utility Function. In SDN-enabled LEO satellite networks, the CPP is a joint optimization of latency and reliability. Different from previous studies, this paper firstly proposes the latency- reliability utility function to comprehensively measure the impact of latency and reliability on LEO satellite network performance. The LRUF of node i is expressed as

$$\eta_i = \alpha \cdot nor_{Ri} + \beta \cdot \frac{1}{nor_{Li}}, \tag{18}$$

where nor_{Ri} and nor_{Li} respectively denote normalized reliability and normalized latency, α and β signify the coefficient of normalized reliability and normalized latency, respectively. Normalized latency is defined as the sum of average propagation latency, average transmission latency and queuing latency of node i, divided by the maximum latency, shown as follows

$$nor_{Li} = \frac{L_i}{L_{max}}, \tag{19}$$

where $L_i = \tau_i^p + \tau_i^t + \tau_i^{que}$ and $L_{max} = dlp + dlt + dlq$.

Controller Placement Problem. CPP is solved in two steps: firstly, the number and location of PSNs are calculated in the network. Secondly, controller placement is performed based on the results of the node and link reliability amendment.

The deployment of PSNs is carried out first, through which INT is performed to improve the overall reliability of the network. Calculate the normalized reliability of all nodes in LEO satellite networks. Then, sort the nodes according to the size of the normalized reliability and select the node with the smallest normalized reliability as the deployment nodes for PSNs. According to Eq. (16) and (17), amend the reliability of the links and nodes as described above. Cycle until all nodes and links in the network are amended.

Based on the results of PSN deployment, with nodes and links reliability amended, controllers are placed on the nodes with the highest LRUF, so that the reliability of the set of controllers reach a large value. Based on the optimal enumeration algorithm, the set of controllers **Con** is placed on the m nodes with the highest LRUFs. The optimization objective is to achieve the maximum sum of LRUF of the set of controllers

$$\max \eta = \sum \eta_i,$$
$$\text{subject to: C1: } \alpha + \beta = 1 \tag{20}$$
$$C2{:}i \in \textbf{Con}$$

where C1 is the constraint for coefficients of latency and reliability and C2 restrains the elements belonging to controller set.

3.2   Programmable Switch-Based Enhanced Controller Placement Algorithm

In this paper, we propose programmable switch-based enhanced controller placement algorithm (PECPA). The detailed steps contain calculating the average reliability under the shortest path from each node to the others in LEO satellite networks, deploying PSNs at the nodes with the lowest reliability, and then deploying PSNs cyclically until the network is covered. Finally, calculate the LRUF of the nodes in the network according to the amendment results, and select the nodes with the largest LRUFs as the set of controllers.

Algorithm 1 Programmable Switch-based Enhanced Controller Placement Algorithm (PECPA)

Input: satellite network topology $G(V,E)$, controller number m

Output: PSN topology V_{INT}, PSN number n_{INT}, controller topology $\mathbf{Con}$

Initialize: $V_{tem} = \varnothing$, $E_{tem} = \varnothing$, $R_i = 0$

Compute R_i of nodes

Sort R_i at descending order

While $\{V_{tem} \cap V_{P4}\} \neq V$ && $E_{tem} \neq E$ do

  Select minimum R_i, add node to V_{INT}, $n_{INT} = n_{INT} + 1$

  Amend reliability of nodes and links, add to V_{tem} E_{tem}

  for $e \in \{E/E_{tem}\}$ do

    If e adjacent nodes involved V_{tem}

      amend the reliability of e, $E_{tem} = E_{tem} \cup \{e\}$

    End if

  End for

End while

Compute η_i of nodes

Select maximum m as $\mathbf{Con}$

return V_{INT}, n_{INT}, $\mathbf{Con}$

---

4 Simulation and Analysis

In this section, we show the results of the comparison between proposed amended and unamended methods. Simulations focus on nodes and links before and after amendment in terms of controller topology, overall network reliability and LRUF.

In the simulation, we simulate a constellation with 66 satellite nodes. The constellation has a total of 6 orbital planes, with 11 satellites on each plane. The altitude of orbit is 780 km, with inclination of 86.4°, and the orbital plane spacing is 30°. To calculate the reliability of the network, the failure probability of the network components is randomly generated from [0.05, 0.2]. To evaluate proposed method, the simulation is conducted in 9 successive time slices, while the interval among slices is 100 s.

Figure 2(a) illustrates the deployment of PSNs and amended nodes in the amended satellite network topology. Where the red diamonds represent the deployment of PSNs,

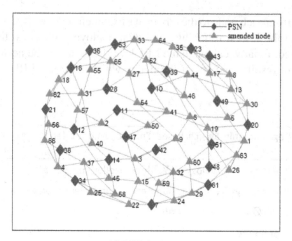

(a) PSN deployment.

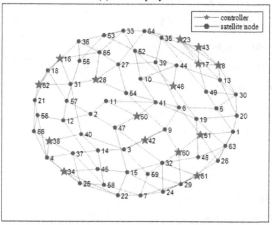

(b) Controller placement before amendment.

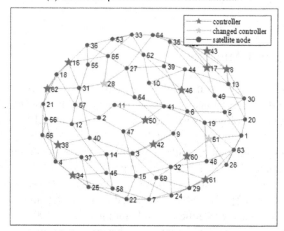

(c) Controller placement after amendment.

Fig. 2. Comparisons of topology.

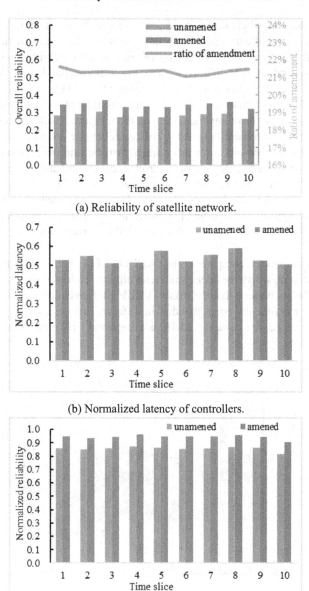

(a) Reliability of satellite network.

(b) Normalized latency of controllers.

(c) Normalized reliability of controllers.

Fig. 3. Comparison of reliability.

the green triangles represent the amended nodes, and the blue lines indicate the amended links. Figure 2(b) represents the placement of the satellite network controllers before amendment, where the pink pentacles are the nodes placed with controllers and the blue dots are satellite switch nodes. Figure 2(c) represents the placement of controllers after reliability amendment, where the changed controllers are highlighted by cyan pentacles. In other words, the set of controllers has been re-selected after the deployment of PSNs.

Figure 3(a) illustrates the comparison of the overall reliability of LEO satellite networks before and after amendment, where the overall reliability of the network is improved by around 20% after amendment compared to that before at the same network topology. Besides, in different time slices, the increased amplitude of overall reliability of the network remains essentially the same over time. Figure 3(c) illustrates the comparison of the normalized reliability of the set of controllers before and after amendment, and the normalized reliability after amendment will be improved around 0.1, which is about 10% normalized reliability of controllers can be enhanced. Additionally, the normalized latency of controller barely changes as shown in Fig. 3(b).

All graphs reflect the effectiveness of proposed amendment method, stably improving the reliability of controllers. Combing Fig. 3(b) and Fig. 3(c), it is indicated that PECPA can increase reliability of controllers in LEO satellite networks without altering those latencies.

Figure 4 illustrates the comparisons of LRUFs of LEO satellite networks before and after the amendment, where the latency and reliability are endowed with the same weight $\alpha = \beta = 0.5$. It can be seen that the overall LRUF after amended increases by about 1 compared to that before amended. Besides, in different time slices, though the satellite topology has changed, the values of LRUF are improved stably, proving the feasibility of PECPA.

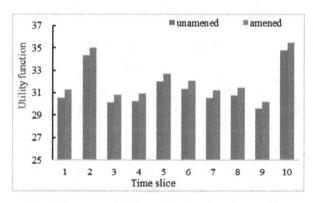

Fig. 4. Latency-reliability utility function of controllers.

5 Conclusions

This paper investigates CPP in LEO satellite networks and proposes node and link reliability amendment based on perceptual enhancement. The LRUF is firstly proposed and PSNs are deployed in the network, achieving the reliability amendment. To maximize LRUF, we propose PECPA and simulation results prove the effectiveness.

As part of our future work, we intend to further investigate the controller and switch migration problem under dynamic topologies considering the combined controller placement problem for both terrestrial and satellite networks.

Acknowledgments. This work is supported by the National Key Research and Development Program (No. 2021YFB2900604), and Young Elite Scientists Sponsorship Program by CAST (No. 2022QNRC001).

Disclosure of Interests. The authors have no competing interests to declare that are relevant to the content of this article.

References

1. Rinaldi, F., et al.: Non-terrestrial networks in 5G beyond: a survey. IEEE Access **8**, 165178–165200 (2020)
2. Liu, J., Shi, Y., Zhao, L., Cao, Y., Sun, W., Kato, N.: Joint placement of controllers and gateways in SDN-enabled 5G-satellite integrated network. IEEE J. Sel. Areas Commun. **36**(2), 221–232 (2018)
3. Chi, C., Yang, L., Huang, Q., Qi, Y.: Optimal placement of multicontroller considering load balance and control delay in software defined satellite network. In: 2022 34th Chinese Control and Decision Conference (CCDC), pp. 2123–2128 (2022)
4. Torkzaban, N., Baras, J.S.: Controller placement in SDN-enabled 5G satellite-terrestrial networks. In: 2021 IEEE Global Communications Conference (GLOBECOM), pp. 1–6 (2021)
5. Luong, D.K., Hu, Y.-F., Li, J.-P., Ali, M.: Metaheuristic approaches to the joint controller and gateway placement in 5G-satellite SDN networks. In: ICC 2020 - 2020 IEEE International Conference on Communications (ICC), pp. 1–6 (2020)
6. Yang, K., Zhang, B., Guo, D.: Controller and gateway partition placement in SDN-enabled integrated satellite-terrestrial network. In: 2019 IEEE International Conference on Communications Workshops (ICC Workshops), pp. 1–6 (2019)
7. Wang, G., Zhao, Y., Huang, J., Yulei, W.: An effective approach to controller placement in software defined wide area networks. IEEE Trans. Netw. Serv. Manage. **15**(1), 344–355 (2018)
8. Li, B., Deng, X., Chen, X., Deng, Y., Yin, J.: MEC-based dynamic controller placement in SD-IoV: a deep reinforcement learning approach. IEEE Trans. Veh. Technol. **71**(9), 10044–10058 (2022)
9. Ji, S., Zhou, D., Sheng, M., Li, J.: Mega satellite constellation system optimization: from a network control structure perspective. IEEE Trans. Wireless Commun. **21**(2), 913–927 (2022)
10. Guo, J., et al.: SDN controller placement in LEO satellite networks based on dynamic topology. In: 2021 IEEE/CIC International Conference on Communications in China (ICCC), pp. 1083–1088 (2021)
11. Guo, J., Yang, L., Rincón, D., Sallent, S., Chen, Q., Liu, X.: Static placement and dynamic assignment of SDN controllers in LEO satellite networks. IEEE Trans. Netw. Serv. Manage. **19**(4), 4975–4988 (2022)
12. Chen, L., Tang, F., Li, X.: Mobility- and load-adaptive controller placement and assignment in LEO satellite networks. In: IEEE INFOCOM 2021 - IEEE Conference on Computer Communications, pp. 1–10 (2021)
13. Liu, Z., Wei, H., Jin, J., Wang, J., Jin, S., Dong, T.: Management of mec service and optimization of mission migration in leo satellite networks. Space-Integr.-Ground Inf. Netw. **3**(3), 72–80 (2022)

14. Donald, G., Shortle John, F., Thompson James, M., Harris Carl, M.: Fundamentals of Queueing Theory. Wiley-Interscience (2008)
15. Yu, B., Chi, X., Liu, X.: Martingale-based bandwidth abstraction and slice instantiation under the end-to-end latency-bounded reliability constraint. IEEE Commun. Lett. **26**(1), 217–221 (2022)
16. Lü, H.R.: Survey on in-band network telemetry. J. Softw. **34**(3870–3890) (2023)

Performance Analysis of Maximum-Likelihood Decoding of Polar Codes

Xiangping Zheng[1,2], Xinyuanmeng Yao[1,2], and Xiao Ma[1,2(✉)]

[1] School of Computer Science and Engineering, Sun Yat-sen University,
Guangzhou 510006, People's Republic of China
`{zhengxp23,yaoxym}@mail2.sysu.edu.cn`
[2] Guangdong Key Laboratory of Information Security Technology, Sun Yat-sen
University, Guangzhou 510006, People's Republic of China
`maxiao@mail.sysu.edu.cn`

Abstract. In space communication, effective channel coding schemes and decoding algorithms are essential for reliable communication. As the standard coding scheme for 5G enhanced Mobile Broadband (eMBB) control channels, polar codes are usually decoded by the successive cancellation list (SCL) decoding algorithm. As the list size increases, the SCL decoding performance is improved and approaches the Maximum Likelihood (ML) decoding performance, while leading to significantly increased complexity. Therefore, analyzing the ML decoding performance is helpful for us to select a suitable list size and hence to trade off the performance and the complexity of the SCL decoding of polar codes. In this paper, we employ the improved union bound and the lower bounding technique based on Bonferroni inequality to evaluate the ML decoding performance of polar codes. The former requires only a truncated weight spectrum and the latter relies only on a subset of the codebook. To calculate the weight spectrum of a polar code, random interleavers are introduced. In contrast, a subset of the codebook can be obtained by performing the SCL decoding algorithm. Simulation results show that the proposed bounding techniques can effectively predict the ML decoding performance of the polar codes and provide guidelines on the choice of the parameters of the SCL decoding. Specifically, the SCL decoding with list size 2 is sufficient to approach the ML decoding performance of the 5G polar codes $[128, 64]$, $[256, 128]$, and $[512, 256]$. By contrast, for the 5G polar code $[128, 16]$ with a 4-bit cyclic redundancy check (CRC), the list size is required to be 8 to obtain near ML decoding performance.

Keywords: Bonferroni inequality · maximum likelihood decoding · polar code · successive cancellation list decoding · truncated weight spectrum

1 Introduction

Channel coding is a key physical layer technology for message transmission in space information networks, including scenarios such as space-to-ground and

Q. Yu (Ed.): SINC 2023, CCIS 2057, pp. 115–127, 2024.
https://doi.org/10.1007/978-981-97-1568-8_10

inter-satellite communications. Therefore, it is necessary to study coding techniques to meet the requirements of different scenarios. Error correction coding techniques such as convolutional codes, Reed-Solomon (RS) codes, Reed-Muller (RM) codes, Turbo codes, and Low-Density Parity-Check (LDPC) codes have been successively introduced into various space communication scenarios. Polar codes, proposed by Arıkan in 2009, are a class of capacity-achieving codes for binary-input output-symmetric discrete-memoryless channels (BIOS-DMCs) [2] and have been adopted as the fifth-generation (5G) standard coding scheme [1]. However, polar codes with the successive cancellation (SC) decoding algorithm [2] does not perform well in the finite code length regime. The belief propagation (BP) decoding algorithm [3] improves the decoding performance but still falls short when compared to the maximum likelihood (ML) decoding algorithm. Later, the successive cancellation list (SCL) decoding algorithm [23,24] was proposed and can achieve ML decoding performance when the list size is sufficiently large. To further enhance the performance of polar codes, the cyclic redundancy check (CRC)-aided SCL decoding algorithm was proposed in [17], referred as to the CA-SCL decoding algorithm. Similar to the CA-SCL decoding algorithm, Wang et al. [26] proposed the parity check (PC)-SCL decoding algorithm.

For the performance evaluation of the SC decoding, the Gaussian approximation (GA) method [25] can be employed. The basic idea of the GA method is to utilize a simplified density evolution algorithm to compute the decoding error probability of each sub-channel and then estimate the decoding error probability of the SC decoding algorithm based on the information bit positions. For the performance evaluation of the SCL decoding, the Monte Carlo simulation method [16] can be used. The Monte Carlo simulation estimates the decoding error probability through a large number of simulation samples, and the accuracy of the estimation depends on the number of simulation samples. However, the Monte Carlo simulation method requires a long time in the high signal-to-noise ratio (SNR) region, and it cannot provide guidelines on the choice of parameters regarding the code design. For the performance evaluation of the ML decoding, the bounding techniques can be employed. As optimal ML decoding achieves the minimum frame error rate (FER), the tight bounds can gain insight into the design of high-performance encoding and decoding algorithms. The union bound is the simplest upper bound. However, it is very loose in the low SNR region. Gallager proposed the first bounding technique (GFBT) [10,20], where the "good region' $\mathcal{R}$ is specified. When the received vector is in the region $\mathcal{R}$, bounding techniques like the union bound are used to characterize the error probability. When the received vector is outside the region $\mathcal{R}$, it is directly considered as a decoding error event. Different choices of this region result in various upper bounds, such as the tangential bound (TB) [5], sphere bound (SB) [11], and tangential-sphere bound (TSB) [9]. However, these upper bounds have high computational complexity due to the numerical integration operation. Ma et al. optimized the design of the region using Hamming distance and proposed two bounding techniques based on the truncated weight spectrum [14] and the geometrical spectrum [15], which reduce the computational complexity by reducing the number of codewords involved. Conventional lower bounds on the ML decoding error probability include the sphere packing bound (SPB) [21]. In 1998, Seguin [19] utilized De Caen's lower bound [8] to propose a lower bound based on pair-wise error probabilities. Behnamfar et al. [4] employed KAT lower bound [13] to analyze the ML decoding performance. Cohen et al. provided a more general lower bound by employing the Cauchy-Schwarz inequality and also proposed a lower bound based on a codeword set with the same Hamming weight [7]. Recently, for polar codes, Shuval et al. proposed an algorithm based on the bit correlations of codewords to compute the triplet-wise error probabilities among codewords and further derived a lower bound based on pair-wise and triplet-wise error probabilities of codewords [22].

This paper employs the improved union bound and the Bonferroni lower bound to predict the ML decoding performance of polar codes. The improved union bound can estimate the ML decoding performance based on the truncated weight spectrum. The computation of the lower bound only requires selecting a codeword subset of polar codes. When the codewords in the selected set have light Hamming weights, a tight Bonferroni lower bound can be calculated using the second-order Bonferroni inequality. We evaluate these two bounding techniques on 5G polar codes. Numerical results demonstrate that the improved union bound and the Bonferroni lower bound based on a codeword set have low computational complexity. Furthermore, the upper and lower bounds are close, allowing for accurate estimation of the ML decoding performance of polar codes and obtaining the minimum list size required for the SCL decoding to approach the ML decoding performance.

2 The Upper and Lower Bounds of Polar Codes

2.1 Polar Codes

System Model. Based on the channel combing and splitting, the channels will show a polarization behavior. When the code length N is sufficiently large, the resulting N sub-channels through channel polarization will be polarized, with the capacity of some sub-channels tending to 1 (noiseless channels) and the capacity of some sub-channels tending to 0 (pure noisy channels). Therefore, it is possible to transmit information bits over reliable sub-channels and frozen bits over unreliable sub-channels to achieve message transmission. Let $\mathcal{P}[N, K, \mathcal{A}]$ denote a polar code with code length N ($N = 2^n, n \geq 1$), code dimension K, and code rate $R = K/N$, where $\mathcal{A} \subseteq \{0, 1, \ldots, N-1\}$ represents the set of indices of K reliable sub-channels for transmitting information bits $\boldsymbol{u}_{\mathcal{A}}$. The remaining $N - K$ sub-channels are indexed by the complementary set $\mathcal{A}^c$ to transmit frozen bits $\boldsymbol{u}_{\mathcal{A}^c}$. Then the codeword $\boldsymbol{c}_0^{N-1} = (c_0, c_1, \ldots, c_{N-1})$ is obtained by $\boldsymbol{c}_0^{N-1} = \boldsymbol{u}_0^{N-1}\mathbf{G}_N$, where $\boldsymbol{u}_0^{N-1} = (\boldsymbol{u}_{\mathcal{A}}, \boldsymbol{u}_{\mathcal{A}^c})$, $\mathbf{G}_N$ represents the generator matrix and $\mathbf{F}^{\otimes n}$ is the Kronecker product of the matrix $\mathbf{F} = \begin{bmatrix} 1 & 0 \\ 1 & 1 \end{bmatrix}$.

Polar codes can also be encoded recursively. Let $c_{m,j}$ represent the j-th codeword with code length 2^m in the m-th stage of the recursion, where $m = 1, 2, \ldots, n$ and $j = 0, 1, \ldots, 2^{n-m} - 1$. Given the codewords $c_{m-1,2j}$ and $c_{m-1,2j+1}$, the codeword $c_{m,j}$ is obtained by

$$c_{m,j} = (c_{m-1,2j} \oplus c_{m-1,2j+1}, c_{m-1,2j+1}), \tag{1}$$

where $c_{0,j} = u_j$ ($0 \leq j \leq N$). Then the polar code of length $N = 2^n$ is represented by the code $c_{n,0}$. Notice that the addition in (1) is the binary addition. Taking a polar code of code length 8 as an example, the encoding process is illustrated in Fig. 1.

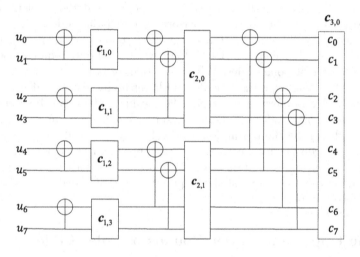

Fig. 1. The encoding process of a polar code with code length 8.

The Weight Spectrum of Polar Codes. For a polar code $\mathcal{P}[N, K, \mathcal{A}]$, the weight spectrum is defined as $A(Y) = \sum_{0 \leq j \leq N} A_j Y^j$, where A_j represents the number of codewords with Hamming weight j and Y is a dummy variable. Since the exact weight spectrum is prohibitively complex when the code size is large, we adopt the average weight spectrum of random interleaved polar (i-polar) codes [6] as an approximation.

The basic idea of i-polar codes [6] is to insert an interleaver Π at the output of each $c_{m-1,2j}$ in (1) during the encoding process. According to [6], the average weight spectrum of random i-polar codes can be recursively computed. When the average weight spectrum of codewords $c_{m-1,2j}$ and $c_{m-1,2j+1}$ denoted as $A_{c_{m-1,2j}}(Y)$ and $A_{c_{m-1,2j+1}}(Y)$, respectively, are available, the average weight spectrum of codeword $c_{m,j}$ is calculated from (2), where the initial condition is shown in (3). When the recursion reaches the stage $m = n$, the average weight spectrum $A_{c_{n,0}}(Y)$ of the random i-polar codes is derived.

$$A_{c_{m,j}}(Y) = \sum_{d_1,d_2} A_{c_{m-1,2j},d_1} A_{c_{m-1,2j+1},d_2} G(d_1, d_2, m), \qquad (2)$$

$$A_{c_{0,j}}(Y) = \begin{cases} 1 + Y, & j \in \mathcal{A} \\ 1, & j \in \mathcal{A}^c \end{cases}, \qquad (3)$$

where,

$$G(d_1, d_2, m) = \sum_{k=\max(0,d_1+d_2-2^{m-1})}^{\min(d_1,d_2)} \frac{\binom{d_2}{k}\binom{2^{m-1}-d_2}{d_1-k}}{\binom{2^{m-1}}{d_1}} Y^{d_1+2d_2-2k}. \qquad (4)$$

Partial Codeword Set of Polar Codes. Decoding errors generally occur when the transmitted codeword is decoded as its neighbor codewords in the Hamming space. To obtain a set of codewords with minimum Hamming weights, an approach based on SCL decoding was proposed by Zhang et al. [27]. Assuming that the all-zero codeword is received, Zhang et al. [27] derived a set of codewords corresponding to the most reliable candidate paths under SCL decoding, which serves as an approximate estimation of the set of codewords with minimum Hamming weights. Moreover, in a more general scenario where the all-zero codeword is transmitted over a noisy channel, non-zero codewords corresponding to the candidate paths are preserved under SCL decoding, resulting in a partial codeword set for polar codes.

2.2   Improved Union Bound

Without loss of generality, assume that the all-zero codeword $c^{(0)}$ is transmitted and modulated by binary phase shift keying (BPSK) result in x_0^{N-1}. Then, the modulated signal vector x_0^{N-1} is transmitted over an additive white Gaussian noise (AWGN) channel, resulting in a received vector $y_0^{N-1} = x_0^{N-1} + w_0^{N-1}$, where w_0^{N-1} represents a N-dimensional Gaussian random vector (each component is a Gaussian random variable with zero mean and variance σ^2). According to [14], the improved union bound on the ML decoding error probability is calculated by

$$\Pr\{E\} \leq \min_{1 \leq d^* \leq N} \left\{ \sum_{d \leq 2d^*} h(A_d) + B(p_b, N, d^* + 1, N) \right\}, \qquad (5)$$

where A_d represents the number of codewords with Hamming weight d,

$$h(A_d) = \min\{p_1(A_d), p_2(A_d)\}, \qquad (6)$$

$$B(p_b, N_l, N_s, N_f) \triangleq \sum_{m=N_s}^{N_f} \binom{N_l}{m} p_b^m (1 - p_b)^{N_l - m}, \tag{7}$$

$$p_b \triangleq Q(1/\sigma), \tag{8}$$

where,

$$p_1(A_d) = A_d Q(\sqrt{d}/\sigma) B(p_b, N - d, 0, d^* - 1), \tag{9}$$

$$p_2(A_d) = (A_d - 1)\left(Q(\sqrt{d}/\sigma) - \frac{1}{2}Q^2(\sqrt{d}/\sigma)\right) B(p_b, N - 2d, 0, d^* - 1) + Q(\sqrt{d}/\sigma), \tag{10}$$

$$Q(x) \triangleq \int_x^{+\infty} \frac{1}{\sqrt{2\pi}} e^{-\frac{z^2}{2}} dz. \tag{11}$$

It is observed that the improved union bound depends only on the truncated weight spectrum where the Hamming weight does not exceed $2d^*$. Given a truncated weight spectrum $A(Y) = \sum_{0 \leq j \leq 2d^*} A_j Y^j$ of the polar code with arbitrary $d^* \geq 0$, the corresponding improved union bound can be calculated. Notice that the d^* can be optimized to obtain a tighter improved union bound.

2.3   Bonferroni Lower Bound

For a polar code $\mathcal{P}[N, K, \mathcal{A}]$, the codebook $\{c^{(i)}, i = 0, 1, \ldots, 2^K - 1\}$ contains 2^K codewords, among which $c^{(0)}$ is the all-zero codeword. In this case, the ML decoding error probability is denoted as $\Pr\{E\} = \Pr\left\{\bigcup_{i=1}^{2^K - 1} \epsilon_{0 \to i}\right\}$, where $\epsilon_{0 \to i}$ represents the event of decoding $c^{(0)}$ to the codeword $c^{(i)}$. Selecting an arbitrary subset $\mathcal{J} \subseteq \{1, \ldots, 2^K - 1\}$ satisfying $|\mathcal{J}| = J$, we have $\Pr\left\{\bigcup_{i=1}^{2^K - 1} \epsilon_{0 \to i}\right\} \geq \Pr\left\{\bigcup_{j \in \mathcal{J}} \epsilon_{0 \to j}\right\}$. According to the second-order Bonferroni inequality [12], we can obtain

$$\Pr\left\{\bigcup_{j \in \mathcal{J}} \epsilon_{0 \to j}\right\} \geq \sum_{i \in \mathcal{J}} \Pr\{\epsilon_{0 \to i}\} - \sum_{i < j \ (i,j \in \mathcal{J})} \Pr\{\epsilon_{0 \to i} \cap \epsilon_{0 \to j}\}. \tag{12}$$

Thus, based on a codeword set, a lower bound on the ML decoding error probability is calculated as follows, denoted as the Bonferroni lower bound.

$$\Pr\{E\} \geq \sum_{i \in \mathcal{J}} \Pr\{\epsilon_{0 \to i}\} - \sum_{i < j \ (i,j \in \mathcal{J})} \Pr\{\epsilon_{0 \to i} \cap \epsilon_{0 \to j}\}, \tag{13}$$

where $\Pr\{\epsilon_{0 \to i}\}$ and $\Pr\{\epsilon_{0 \to i} \cap \epsilon_{0 \to j}\}$ are computed through finite integration [18] by

$$\Pr\{\epsilon_{0 \to i}\} = \frac{1}{\pi} \int_0^{\pi/2} \exp\left[-\frac{W_H(c^{(i)})}{\sigma^2 \sin^2 \theta}\right] d\theta, \tag{14}$$

$$\Pr\{\epsilon_{0\to i} \cap \epsilon_{0\to j}\} = \psi\left(\frac{\sqrt{W_H\left(\boldsymbol{c}^{(i)}\right)}}{\sigma}, \frac{\sqrt{W_H\left(\boldsymbol{c}^{(j)}\right)}}{\sigma}\right), \tag{15}$$

where $W_H(\boldsymbol{c}^{(i)})$ is the Hamming weight of the codeword $\boldsymbol{c}^{(i)}$,

$$\begin{aligned}
\Psi(x,y) = {} & \frac{1}{2\pi}\int_0^{\frac{\pi}{2}-\tan^{-1}\left(\frac{y}{x}\right)} \frac{\sqrt{1-\rho_{ij}}}{1-\rho_{ij}\sin 2\theta}\exp\left[-\frac{x^2(1-\rho_{ij}\sin 2\theta)}{2(1-\rho_{ij}^2)\sin^2\theta}\right]d\theta \\
& + \frac{1}{2\pi}\int_0^{\tan^{-1}\left(\frac{y}{x}\right)} \frac{\sqrt{1-\rho_{ij}}}{1-\rho_{ij}\sin 2\theta}\exp\left[-\frac{y^2(1-\rho_{ij}\sin 2\theta)}{2(1-\rho_{ij}^2)\sin^2\theta}\right]d\theta
\end{aligned} \tag{16}$$

where ρ_{ij} is related to the Hamming distance $w_H\left(\boldsymbol{c}^{(i)}-\boldsymbol{c}^{(j)}\right)$ between the codeword $\boldsymbol{c}^{(i)}$ and the codeword $\boldsymbol{c}^{(j)}$ and can be calculated by

$$\rho_{ij} = \frac{w_H\left(\boldsymbol{c}^{(i)}\right) + w_H\left(\boldsymbol{c}^{(j)}\right) - w_H\left(\boldsymbol{c}^{(i)}-\boldsymbol{c}^{(j)}\right)}{2\sqrt{w_H\left(\boldsymbol{c}^{(i)}\right) w_H\left(\boldsymbol{c}^{(j)}\right)}}. \tag{17}$$

2.4   Performance Analysis of Polar Codes

Based on the average weight spectrum and the codeword set, the improved union bound can be evaluated from (5), and the Bonferroni lower bound can be calculated using (13), thus providing an estimation of the ML decoding performance of polar codes. In the next section, we simulate the 5G polar codes over BPSK-AWGN channels and estimate the ML decoding performance using these two bounding techniques.

3   Numerical Results

3.1   The Performance of 5G Polar Codes

In this subsection, we present numerical results of the upper and lower bounds for estimating the ML decoding performance of 5G polar codes $\mathcal{P}[128,64]$, $\mathcal{P}[256,128]$, and $\mathcal{P}[512,256]$.

We present the optimal parameter d^* of the truncated weight spectrum required in the improved union bound for three 5G polar codes in Fig. 2, from which we see that d^* for $\mathcal{P}[128,64]$, $\mathcal{P}[256,128]$, and $\mathcal{P}[512,256]$ are found to be 45, 68 and 104, respectively, at 3.0 dB. It is also observed that d^* decreases as the SNR increases as expected, since the main errors occur within a smaller corresponding Hamming sphere in the higher SNR region.

Based on the average weight spectrum of the ensemble of i-polar codes, the average number of codewords with minimum Hamming weights can be derived and serves as the list size of the SCL decoding, resulting in a codeword set with

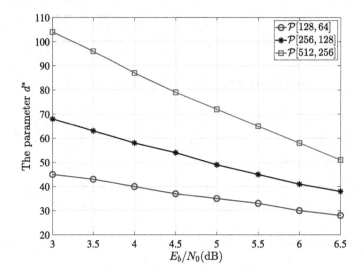

Fig. 2. The optimal parameter d^* required in the improved union bound for 5G polar codes.

light Hamming weights. It is worth noting that to obtain a tight Bonferroni lower bound, different sets of codewords can be selected at different SNRs. Figure 3 presents the distribution of Hamming distances between codewords required in Bonferroni lower bound for 5G polar codes. Notice that from (13), the computational complexity of the Bonferroni lower bound mainly stems from the integration calculation of $\Pr\{\epsilon_{0\to i} \cap \epsilon_{0\to j}\}$. When the Hamming weights of the codewords are the same, the calculation of $\Pr\{\epsilon_{0\to i} \cap \epsilon_{0\to j}\}$ mainly depends on the Hamming distances between codewords. As shown in Fig. 3, it is observed that when selecting 275 codewords with Hamming weight 8 for the 5G polar code $\mathcal{P}[128,64]$, the Hamming distances between codewords are classified into three categories, 8, 12, and 16. Similarly, when selecting 96 codewords with Hamming weight 8 for the 5G polar code $\mathcal{P}[256,128]$, the Hamming distances between codewords are classified into two categories, 8 and 16. For the 5G polar code $\mathcal{P}[512,256]$, when selecting 64 codewords with Hamming weight 8, the Hamming distances between codewords are all 16, indicating that the distribution of Hamming distances between codewords of the selected set exhibits sparsity. Thus, the number of calculations for $\Pr\{\epsilon_{0\to i} \cap \epsilon_{0\to j}\}$ is significantly reduced, resulting in low computational complexity of the Bonferroni lower bound. For example, for the 5G polar code $\mathcal{P}[128,64]$, only three types of $\Pr\{\epsilon_{0\to i} \cap \epsilon_{0\to j}\}$ need to be calculated.

We present the performance of three 5G polar codes under SCL decoding with list size $L = 1, 2$ along with the union bound, the improved union bound, and the Bonferroni lower bound, as shown in Figs. 4, 5 and 6, from which we see that the SCL decoding algorithm with $L = 2$ approaches the ML decoding performance. Since the performance gain of the SCL decoding with $L = 4$ is marginal compared

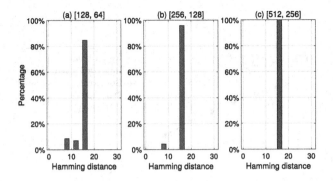

Fig. 3. The distribution of Hamming distances required in Bonferroni lower bound for 5G polar codes.

with the SCL decoding with $L = 2$ and the complexity significantly increases, $L = 2$ is sufficient for 5G polar codes $\mathcal{P}[128, 64]$, $\mathcal{P}[256, 128]$, and $\mathcal{P}[512, 256]$. Without the bounding techniques, it is required to increase the list size for the performance evaluation of the SCL decoding. It is also observed that the improved union bound and Bonferroni lower bound are close in the high SNR region, allowing us to estimate the FER performance of polar codes under ML decoding.

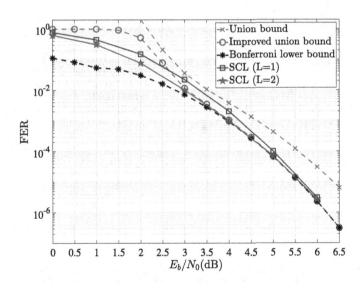

Fig. 4. The performance of the 5G polar code $\mathcal{P}[128, 64]$.

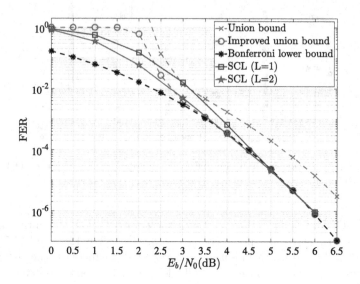

Fig. 5. The performance of the 5G polar code $\mathcal{P}[256, 128]$.

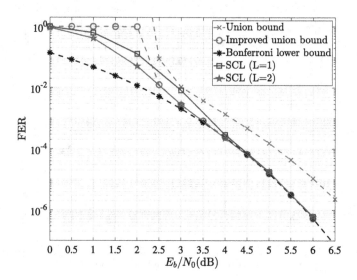

Fig. 6. The performance of the 5G polar code $\mathcal{P}[512, 256]$.

3.2   The Performance of 5G Polar Codes Aided by CRCs

In practical applications, polar codes are typically concatenated with CRCs. It becomes difficult to calculate the weight spectrum, thus simple upper bounds are not available. However, a codeword subset of polar codes can always be obtained through CA-SCL decoding, and the Bonferroni lower bound can be employed. When the CA-SCL decoding performance approaches the Bonferroni lower bound, the lower bound technique can be utilized to guide the selection of list size. In this subsection, we simulate the 5G polar code $\mathcal{P}[128, 16]$ under the CA-SCL decoding algorithm with a 4-bit CRC.

Figure 7 presents the SCL decoding performance of the 5G polar code $\mathcal{P}[128, 16]$ with a 4-bit CRC, along with the Bonferroni lower bound, from which it is observed that as the list size increases, the CA-SCL decoding performance gradually improves, and has already approached the ML decoding performance at 6.0 dB with list size 8, indicating list size 8 is sufficient for the 5G polar code $\mathcal{P}[128, 16]$ in the high SNR region. Note that compared with the polar codes without CRC concatenation, the list size required for the CA-SCL algorithm to approach the ML decoding performance is generally larger. For example, the list size of 2 is sufficient for the 5G polar code $\mathcal{P}[128, 64]$, while the list size for the 5G polar code $\mathcal{P}[128, 16]$ with a 4-bit CRC is set to 8.

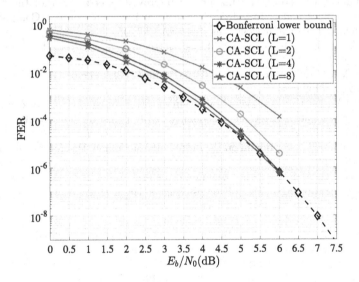

Fig. 7. Performance of the 5G polar code $\mathcal{P}[128, 16]$ with a 4-bit CRC.

4   Conclusion

This paper is concerned with the bounding techniques to estimate the ML decoding performance of polar codes, including the improved union bound and the

Bonferroni lower bound based on a codeword subset of polar codes. Numerical results demonstrate that the upper and lower bounds are close, resulting in an effective estimation of the ML decoding performance of polar codes and providing the guidance of the list size for the SCL decoding algorithm. Specifically, for 5G polar codes [128, 64], [256, 128], and [512, 256], list size 2 is sufficient for the SCL decoding algorithm. For the 5G polar code [128,16] with a 4-bit CRC, list size 8 is sufficient.

Acknowledgments. This work was supported in part by the National Key R&D Program of China (No. 2021YFA1000500).

References

1. 3GPP: Multiplexing and channel coding V.15.3.0. Sophia Antipolis, France, Rep. 38.212 (2018)
2. Arıkan, E.: Channel polarization: a method for constructing capacity-achieving codes. In: IEEE International Symposium on Information Theory, pp. 1173–1177 (2008)
3. Arıkan, E.: A performance comparison of polar codes and Reed-Muller codes. IEEE Commun. Lett. **12**(6), 447–449 (2008)
4. Behnamfar, F., Alajaji, F., Linder, T.: Improved lower bounds for the error rate of linear block codes. In: 43rd Allerton Conference on Control, Computing and Communications, Monticello, Illinois, USA (2005)
5. Berlekamp, E.: The technology of error-correcting codes. Proc. IEEE **68**(5), 564–593 (1980)
6. Chiu, M.C.: Interleaved polar (I-Polar) codes. IEEE Trans. Inf. Theory **66**(4), 2430–2442 (2020)
7. Cohen, A., Merhav, N.: Lower bounds on the error probability of block codes based on improvements on De Caen's inequality. IEEE Trans. Inf. Theory **50**(2), 290–310 (2004)
8. De Caen, D.: A lower bound on the probability of a union. Discret. Math. **169**(1–3), 217–220 (1997)
9. Divsalar, D.: A simple tight bound on error probability of block codes with application to turbo codes. TMO Progress Report **19**, 42–139 (1999)
10. Gallager, R.: Low-density parity-check codes. IRE Trans. Inf. Theory **8**(1), 21–28 (1962)
11. Herzberg, H., Poltyrev, G.: Techniques of bounding the probability of decoding error for block coded modulation structures. IEEE Trans. Inf. Theory **40**(3), 903–911 (1994)
12. Hoppe, F.M.: Iterating bonferroni bounds (1985)
13. Kuai, H., Alajaji, F., Takahara, G.: Tight error bounds for nonuniform signaling over AWGN channels. IEEE Trans. Inf. Theory **46**(7), 2712–2718 (2000)
14. Ma, X., Liu, J., Bai, B.: New techniques for upper-bounding the ML decoding performance of binary linear codes. IEEE Trans. Commun. **61**(3), 842–851 (2013)
15. Ma, X., Liu, J., Zhuang, Q., Bai, B.: New geometrical spectra of linear codes with applications to performance analysis. In: IEEE Information Theory Workshop, pp. 1–5 (2013)
16. Mooney, C.Z.: Monte Carlo Simulation. Sage, Thousand Oaks (1997)

17. Niu, K., Chen, K.: CRC-aided decoding of polar codes. IEEE Commun. Lett. **16**(10), 1668–1671 (2012)
18. Sason, I., Shamai, S., et al.: Performance analysis of linear codes under maximum-likelihood decoding: a tutorial. Found. Trends® Commun. Inf. Theory **3**(1–2), 1–222 (2006)
19. Seguin, G.: A lower bound on the error probability for signals in white Gaussian noise. IEEE Trans. Inf. Theory **44**(7), 3168–3175 (1998)
20. Shamai, S., Sason, I.: Variations on the Gallager bounds, connections, and applications. IEEE Trans. Inf. Theory **48**(12), 3029–3051 (2002)
21. Shannon, C.E.: Probability of error for optimal codes in a Gaussian channel. Bell Syst. Tech. J. **38**(3), 611–656 (1959)
22. Shuval, B., Tal, I.: A lower bound on the probability of error of polar codes over BMS channels. IEEE Trans. Inf. Theory **65**(4), 2021–2045 (2019)
23. Tal, I., Vardy, A.: List decoding of polar codes. In: IEEE International Symposium on Information Theory Proceedings, pp. 1–5 (2011)
24. Tal, I., Vardy, A.: List decoding of polar codes. IEEE Trans. Inf. Theory **61**(5), 2213–2226 (2015)
25. Trifonov, P.: Efficient design and decoding of polar codes. IEEE Trans. Commun. **60**(11), 3221–3227 (2012)
26. Wang, T., Qu, D., Jiang, T.: Parity-check-concatenated polar codes. IEEE Commun. Lett. **20**(12), 2342–2345 (2016)
27. Zhang, Q., Liu, A., Pan, X., Pan, K.: CRC code design for list decoding of polar codes. IEEE Commun. Lett. **21**(6), 1229–1232 (2017)

An Efficient Transmission-Reception Scheme for LoRa-Based Uplink Satellite IoT Communications

Zhongyang Yu[✉], Mengmeng Xu, Yuanyuan Wang, and Jixun Gao

School of Computer Science, Henan University of Engineering,
Zhengzhou, People's Republic of China
xd_yzy2013@sina.com, jsj@haue.edu.cn
http://jsj.haue.edu.cn

Abstract. Due to the impact of the superposition of noises and interferences caused by the users using the long range (LoRa) modulation with large spreading factors (SFs) for uplink satellite IoT communications, the other users using the LoRa with relatively small SFs are hard to detect without the aid of the power allocation and interference cancellation. To this end, we propose a simple but efficient transmission-reception scheme including selective repetition transmission and superposition reception. In specific, the selective repetition can support two cases of full repetition and partial repetition in order to achieve a good trade-off between detection performance and transmission rate, and the superposition reception can mitigate the negative effect of the superposition of the noises and interferences. Simulation results validate the efficiency of the proposed scheme.

Keywords: LoRa modulation · Repetition transmission · Superposition reception · Satellite IoT

1 Introduction

Satellite Internet of Things (IoT) networks will be capable of solving the issue of massive users' connectivity especially in remote areas [1,2]. Unlike terrestrial access networks such as long range (LoRa) technology with multiple gateways, satellite-based networks perhaps need to consider joint orthogonal "time-frequency-code" division multiple access technology such as the classical time division multiple access (TDMA), frequency division multiple access (FDMA), and code division multiple access (CDMA). For this reason, we try to design a joint "frequency-code" orthogonal multiple access scheme, combining pseudo-noise (PN) or Walsh coding and LoRa modulation with different spreading factors (SFs). First of all, this letter will deal with the problem of detecting superposed LoRa signals without the use of the power allocation and interference cancellation strategy.

For the LoRa modulation, it can be described as the frequency shift chirp modulation (FSCM) [3], where the initial frequency shift carries the user information unlike the pure chirp spread spectrum (CSS) modulation [4] but similar to a kind of high-order CSS. In [5], the authors provided a detailed analysis of LoRa waveform property and its spectral, where they proved that the LoRa waveforms are asymptotically orthogonal for

Q. Yu (Ed.): SINC 2023, CCIS 2057, pp. 128–138, 2024.
https://doi.org/10.1007/978-981-97-1568-8_11

large SFs. In other words, those with relatively small SFs exhibit weak orthogonality. Considering two LoRa signals simultaneously received with the same SFs, Ref. [6] showed that the reception can be successful as long as one signal is received at least 6 dB above the other while Ref. [7] presented a different result and further indicated that packet loss would occur in the packets modulated with different SFs if the interference is strong enough. However, these results were obtained under the noise-free environment. Indeed, only the LoRa signal with a high SF is possible to detected from the superposed LoRa signals in the presence of Gaussian noises [8], which potentially proves the LoRa waveform property derived in [5]. For the superposed LoRa signals with the same SFs, Temim *et al.* addressed the corresponding reception based on the well-known successive interference cancellation (SIC) algorithm [9]. Furthermore, Noreen *et al.* investigated the capture effect of the LoRa based system with different SFs and its combination with the SIC algorithm [10]. In order to maximize energy efficiency of the system, Su *et al.* studied the corresponding energy efficient resource allocation which includes the power allocation [11]. However, the performance of the LoRa based systems in these researches is determined by the energy threshold selection in the power allocation and SIC strategy regardless of using the same SFs or different SFs, where in this case the strategy seems complicated.

In this paper, we deal with the issue of uplink multiuser transmission based on the LoRa modulation with different SFs without the use of the power allocation and SIC strategy. At the transmitter, we propose a selective repetition transmission scheme for the users with small SFs. Accordingly, a superposition reception scheme is provided to detect those users at the receiver. Taking two-user and three-user simultaneous trans-missions for examples, simulation results demonstrate that the proposed scheme can perform well especially for the multiuser transmission with far SFs. Meanwhile, fading performance evaluation and massive-user transmission design are also given.

2   System Model

Consider an uplink (satellite) IoT system with N users sharing an available bandwidth B, where each user adopts the LoRa modulation with an unique SF. In other words, $SF_1 \sim SF_N$ are assigned in turn to *User* $1 \sim N$, restricting $SF_1 < SF_2 < \cdots < SF_N$ without loss of generality. Under such conditions, the l-th received signal, taken at the chirp rate B, can be expressed as

$$y_N(l,k) = \sum_{m=1}^{N} s_m(l,k)\, e^{j(2\pi f_d^m k/B + \theta_m)} + n_m(l,k), \qquad (1)$$

where f_m^m and θ_m are the frequency offset and phase offset, respectively, k is the chirp index, whose specific range will be given latter, $s_m(l,k)$ is the LoRa signal with the average energy $E_s = \mathbb{E}\{|s_m(l,k)|^2\} = 1$, whose specific form will be given in the next section, and $n_m(l,k)$ is the additive white Gaussian noise (AWGN) with zero mean and variance σ_m^2.

Since the noncoherent detection will be used, the impact of the phase offset is thus not considered. Furthermore, we will consider in simulation the frequency offset $|f_d^m| < 0.5B/2^{SF_m}$ according to the interval between the two adjacent chirps.

3   Proposed Scheme

In general, the users using the LoRa modulation with small SFs are susceptible to the impact of the noises and interferences from other users using the LoRa with relatively large SFs. This is because the LoRa modulation with small SFs is unsatisfactory in terms of both noise-immune ability and orthogonality [5]. If not consider the power allocation and SIC strategy, our selective repetition transmission and superposition reception scheme can be selected.

3.1   Selective Repetition Transmission

To start with, we define the chirp number $M_i = 2^{SF_i}$ for $User\ i$, followed by a varying number of repeats K_i ($i \neq N$). Clearly, $K_i \leq (M_N - M_i)/M_i$ in terms of the chirp number $M_N = 2^{SF_N}$ for $User\ N$. Furthermore, we denote by $K_i = (M_N - M_i)/M_i$ the number of full repetition and $K_i < (M_N - M_i)/M_i$ the number of partial repetition, designated as K_i^{fr} and K_i^{pr} respectively.

Based on repeating the original symbols generated by the LoRa modulation with small SFs, we put forward a selective repetition transmission scheme, as shown in Fig. 1 with (a) no repetition case, (b) full repetition case, and (c) partial repetition case. For the *no repetition case*, there are M_N/M_i original LoRa symbols to be transmitted. For the *full repetition case*, it carries only one original symbol along with the same K_i^f symbols obtained by repeating this original symbol. For the *partial repetition case*, it supports one current original symbol and its K_i^p repeated symbols as well as other symbols and their $\tilde{K}_i^p$ repeated symbols. Surprisingly, for $User\ i$, the latter two cases can mitigate the other users' interferences in different degrees at the expense of some transmission rate, compared to the case (a). However, it is known from the existing literatures [9–11] that the power allocation and SIC strategy is required for the case (a).

For $User\ i$, the original LoRa symbol can be represented as [3]

$$s_i^o(l,k) = \frac{e^{j2\pi\left[\left(d_i^i+\frac{k}{2}\right) \bmod M_i\right]\frac{k}{M_i}}}{\sqrt{M_i}}, \quad k \in \{0,1,\dots,M_i-1\}, \tag{2}$$

where $d_i^i \in \{0,1,\dots,M_i-1\}$ is the decimal number determined by SF_i binary digits, the term *mod* denotes the modulo operator.

In the following, we will explain the three cases above.

1) No repetition case: The modulation signal $s_i(l,k)$ in (1) consists of M_N/M_i different original symbols, each having similar form of $s_i^o(l,k)$ in (2)

$$s_i(l,k) = \left[\underbrace{s_i^o(l,k)}_{\text{current symbol}} \quad \overbrace{s_i^o(l+1,k)\cdots}^{(M_N/M_i-1)\ \text{other symbols}} \right]. \tag{3}$$

The corresponding transmission rate is thus $R_{b,i}^{nr} = \frac{B \cdot SF_i}{M_i}$.

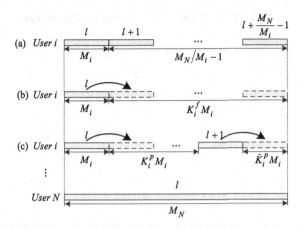

Fig. 1. Schematic diagram of the proposed transmission scheme for *User i*, where the solid box and the dashed box denote the original LoRa symbol and its repeated version, respectively.

2) Full repetition case: The current symbol $s_i^o(l,k)$ is repeated K_i^f times, yielding a new version of the modulation signal $s_i(l,k)$:

$$s_i(l,k) = \left[\underbrace{s_i^o(l,k)}_{\text{current symbol}} \overbrace{s_i^o(l,k)\cdots s_i^o(l,k)}^{K_i^f \text{ repeated symbols}} \right], \tag{4}$$

where these repeated symbols will be fully explored during the noncoherent detection. But the corresponding transmission rate is reduced by a factor of (K_i^f+1) denoted by $R_{b,i}^{fr} = \frac{R_{b,i}^{nr}}{K_i^f+1}$, compared to the no repetition case.

3) Partial repetition case: The current symbol $s_i^o(l,k)$ and other symbols such as $s_i^o(l+1,k)$ are repeated K_i^p times and $\tilde{K}_i^p$ times, respectively, obtaining another new version of the modulation signal $s_i(l,k)$:

$$s_i(l,k) =$$

$$\left[\underbrace{s_i^o(l,k)}_{\text{current symbol}} \overbrace{s_i^o(l,k)\cdots}^{K_i^p \text{ repeated symbols}} \underbrace{s_i^o(l+1,k)}_{\text{other symbols}} \overbrace{\cdots}^{\tilde{K}_i^p \text{ repeated symbols}} \right], \tag{5}$$

where partial repeated symbols are utilized, which equivalently improves the loss of transmission rate against the full repetition case. In this sense, the corresponding transmission rate becomes $R_{b,i}^{pr} = R_{b,i}^{fr} \cdot \frac{K_i^f+1}{K_i^p+1}$.

3.2 Superposition Reception

For the full repetition case or partial repetition case, we will consider the *superposition* of the current symbol and its repeated symbols, as revealed in Fig. 2, in order to enhance

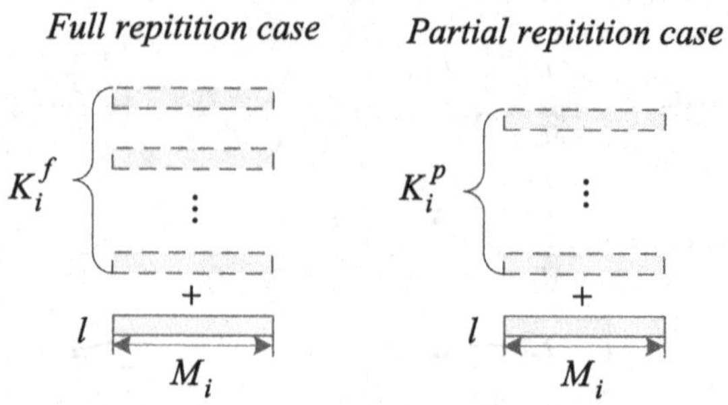

Full repitition case   *Partial repitition case*

Fig. 2. Schematic diagram of the corresponding reception scheme for *User i*.

the ability of anti-interference and anti-noise especially for the users using the LoRa modulation with fairly small SFs. For simplicity, we will omit the frequency offset and phase offset in the following discussion.

For *User i*, other users are considered as interferences such that (1) can be rewritten as

$$y_N(l, k) = s_i(l, k) + \underbrace{\sum_{m=1, m\neq i}^{N} s_m(l, k)}_{\text{other user's interferences}} + \underbrace{\tilde{n}(l, k)}_{\text{superposed noises}}, \tag{6}$$

where $\tilde{n}(l, k) \triangleq \sum_{m=1}^{N} n_m(l, k)$.

1) No repetition case: We consider (2) and (3) into (6), and multiply by a down chirp with the form of $e^{-j\pi k^2/M_i}$ such that a dechirped signal can be derived as

$$\tilde{y}_N^{nr}(l, k) = \frac{1}{\sqrt{M_i}} e^{j2\pi \frac{d_i^i}{M_i} k} + s_I(l, k) + \tilde{n}(l, k), \tag{7}$$

where $s_I(l, k) \triangleq \sum_{m=1, m\neq i}^{N} s_m(l, k) e^{-j\pi k^2/M_i}$ is the superposition of $N-1$ LoRa-like signals, each having a wide-band spectrum with low spectral density [7], $\tilde{\tilde{n}}(l, k) \triangleq \tilde{n}(l, k) e^{-j\pi k^2/M_i}$ is the rotation version of $\tilde{n}(l, k)$.

Then, we can compute (7) via the discrete Fourier transform (DFT) using M_i orthogonal chirps, given by

$$R_l^{nr}(q) = \delta(q - d_l^i) + S_I(q) + N(q), q = 0, 1, \ldots, M_i - 1, \tag{8}$$

where $S_I(q)$ is the DFT of $s_I(l, k)$, $N(q)$ is the DFT of $\tilde{\tilde{n}}(l, k)$. One can observe that it is hard to detect the desired symbol d_l^i from the signal $R_l^{nr}(q)$ due to the effect of the interference $S_I(q)$ and the noise $N(q)$.

2) Full repetition case: Similarly, we have after some calculations

$$\tilde{y}_N^{fr}(l, k) = \frac{K_i^f + 1}{\sqrt{M_i}} e^{j2\pi \frac{d_i^i}{M_i} k} + \sum_{v=0}^{K_i^f} s_I(l+v, k) + \tilde{n}(l+v, k), \tag{9}$$

En Efficient Transmission-Reception Scheme

and

$$R_l^{fr}(q) = \left(K_i^f + 1\right)\delta\left(q - d_l^i\right) + \sum_{v=0}^{K_i^f}[S_I(q) + N(q)]\,e^{j2\pi qv},$$
$$q = 0, 1, \ldots, M_i - 1. \tag{10}$$

Actually, one can find from (10) that the superposition of the current symbol and its repeated symbols can mitigate the negative impact of both interference and noise from other users, which helps to capture the expected symbol d_l^i.

3) Partial repetition case: Like the full repetition case, we also have

$$\tilde{y}_N^{pr}(l, k) = \frac{K_i^p + 1}{\sqrt{M_i}}e^{j2\pi\frac{d_l^i}{M_i}k} + \sum_{v=0}^{K_i^p}s_I(l + v, k) + \tilde{\tilde{n}}(l + v, k), \tag{11}$$

and

$$R_l^{pr}(q) = (K_i^p + 1)\delta\left(q - d_l^i\right) + \sum_{v=0}^{K_i^p}[S_I(q) + N(q)]\,e^{j2\pi qv},$$
$$q = 0, 1, \ldots, M_i - 1. \tag{12}$$

Due to $K_i^p < K_i^f$, the partial superposition reception can guarantee a good compromise between the detection performance and the transmission rate.

Finally, the transmitted symbol d_l^i can be estimated after taking the amplitude of (8), (10), or (12) and looking for the location of the maximum [3,5,12]. Obviously, such noncoherent detection is not affected by the phase offset in (1) owing to taking the *amplitude* rather than the *real* part.

With regard to *User N* assigned to the largest SF, the corresponding anti-noise ability and orthogonality are the strongest such that the transmitted symbol can easily be detected, which will be verified in the latter simulation.

4   Simulation Results

In this section, we will evaluate the performance of the proposed scheme to detect simultaneously received LoRa signals with different SFs under the AWGN channel. The basic parameters are listed in Table 1.

4.1   Two Users with Near or Far SFs

1) The case of two users with near SFs: Assume that *User* 1 and *User* 2 use $SF_1 = 7$ and $SF_2 = 9$, respectively. Figure 3 presents the detection performance of the two-user simultaneous transmission using $SF_1 = 7$ and $SF_2 = 9$, with (a) no frequency offsets and (b) the default frequency offsets.

As observed from Fig. 3(a), for *User* 1 the corresponding performance improvement seems remarkable when considering the full repetition case ($K_1^f = 4$) or the partial repetition case ($K_1^p = 1$), compared to the no repetition case. In Specific, the performance

Table 1. Basic parameters

Bandwidth (kHz)	$B = 125$
Spreading factor	$SF = 7, 9, 12$
Chirp number	$M = 128, 512, 4096$
Number of users	$N = 2, 3$

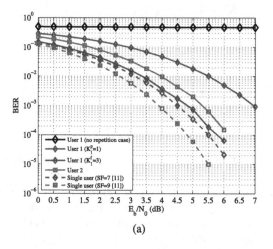

(a)

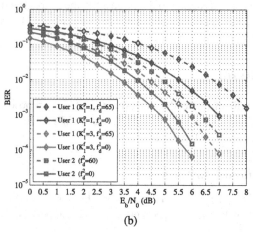

(b)

Fig. 3. Detection performance of the two-user simultaneous transmission using $SF_1 = 7$ and $SF_2 = 9$ with (a) no frequency offsets and (b) the default frequency offsets.

of $K_1^f = 4$ can achieve the performance near to the single user transmission with $SF = 7$ [12], but at the expense of the transmission rate (where $R_{b,1}^{fr} \approx 1709$ bps and $R_{b,1}^{nr} \approx 6835$ bps). Although the performance loss of $K_1^p = 1$ approaches 2.5 dB at the BER of 10^{-3}, the corresponding transmission rate can be increased to $R_{b,1}^{pr} \approx 3418$ bps. Moreover, the performance loss of *User* 2 is near 1.3 dB compared to the single user transmission with $SF = 9$ [12]. On the other hand, the two users are still capable of being detected well under small frequency offsets (and random phase offsets), which has reflected in Fig. 3(b).

2) The case of two users with far SFs: Assume that *User* 1 and *User* 2 adopt $SF_1 = 7$ and $SF_2 = 12$, respectively. For *User* 1, we consider the no repetition case with 32 symbols transmitted, the full repetition case with $K_1^f = 31$, and the partial repetition case with $K_1^p = 7, 15$. Figure 4 presents the corresponding detection performance.

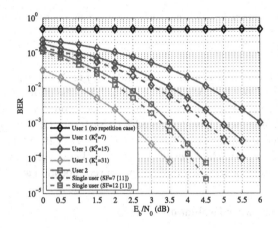

Fig. 4. Detection performance of the two-user simultaneous transmission with $SF_1 = 7$ and $SF_2 = 12$.

It is found from Fig. 4 that the performance of $K_1^f = 31$ can achieve the best performance thanks to the superposition of up to 31 repeated symbols, but resulting in a great sacrifice of the transmission rate (note that $R_{b,1}^{fr} \approx 214$ bps and $R_{b,1}^{nr} \approx 6835$ bps). Conversely, the performance losses of $K_1^p = 15$ and $K_1^p = 7$ can approach 0.5 dB and 1.5 dB against the single user transmission with $SF = 7$ [12], while the corresponding transmission rates are enhanced to 428 bps and 856 bps, respectively. Moreover, considering the single user transmission with $SF = 12$ [12], the performance loss of *User* 2 is actually negligible, which proves the previous discussion on the user with the largest SF.

4.2 Three Users with Different SFs

Suppose that *User* 1, *User* 2, and *User* 3 utilize $SF_1 = 7$, $SF_2 = 9$, and $SF_3 = 12$, respectively. The detection performance of the three-user simultaneous transmission based on the proposed scheme is provided in Fig. 5.

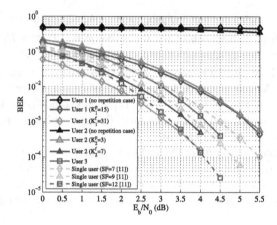

Fig. 5. Detection performance of the three-user simultaneous transmission using $SF_1 = 7$, $SF_2 = 9$, and $SF_3 = 12$

In addition to the similar results to Fig. 3(a) and Fig. 4, one can also find that the performance loss of *User* 3 becomes larger compared to the performance of the single user with $SF = 12$ [11] due to the double interferences from *User* 1 and *User* 2. Also, increased performance losses have happened for *User* 1 and *User* 2, regardless of considering the full repetition case ($K_1^f = 31$ and $K_2^f = 7$) or the partial repetition case ($K_1^p = 15$ and $K_2^p = 3$). These results allow us to know the limit of the proposed scheme, though there is no need to consider the power allocation and SIC strategy.

4.3 Fading Performance Evaluation

Taking the two-user transmission using far SFs for an example, we will evaluate the performance of the proposed scheme over a frequency-flat Rician fading channel with different K factors $K = 5$ dB, $K = 10$ dB and $K = 15$ dB, where *User* 1 and *User* 2 adopt $SF_1 = 7$ and $SF_2 = 12$, respectively. For *User* 1, the full repetition case with $K_1^f = 31$ is considered. The corresponding performance curve is depicted in Fig. 6.

We note that for *User* 2 with $SF_2 = 12$ the corresponding performance under the Rician channel with $K = 10$ dB and $K = 15$ dB is almost close to that under the AWGN channel (from Fig. 4), while the performance with $K = 5$ dB starts to become poor. For *User* 1 with $SF_2 = 7$ and $K_1^f = 31$, the overall performance under the Rician channel with different K factors seems much worse than that under the AWGN channel (from Fig. 4). In other words, these results tells us to know that the impact of fading on the user assigned to the small SF is greater than that on the user assigned to the large SF.

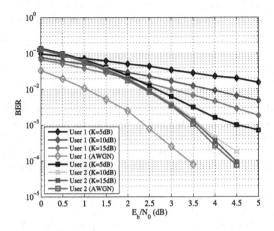

Fig. 6. Detection performance of the two-user simultaneous transmission using $SF_1 = 7$ and $SF_3 = 12$ over the frequency-flat Rician fading channel

4.4 Massive-User Transmission Design

As mentioned in the introduction, we will consider the combination of a PN (or Walsh) code with L offsets (or rows) and a LoRa modulation with N SFs (typically $L = 512$ and $N = 2$). In general, $U = L \times N$ users are divided into L groups, each group having an unique offset. All N users in each group are modulated by the N-SF LoRa modulators, respectively. Based on the match-filtering method [13, 14] and our proposed scheme, these U users may be successfully detected even in the present of large frequency offsets. Further research will be carried out in our extended work.

5 Conclusion

In the letter we have shown that, for uplink satellite IoT communications with the LoRa modulation the user assigned the largest SF can easily be captured even when simultaneously transmitting several other users assigned small SFs. To detect these users, we designed a selective repetition transmission and superposition reception scheme. By considering the two cases of the full repetition case and partial repetition case, a good tradeoff can be achieved between the detection performance and transmission rate. Simulation results indicate that, the proposed scheme for two-user transmission with near SFs can provide higher transmission rates but unsatisfactory performances, compared to that for two-user transmission with far SFs; and more users' transmission will reduce the overall performance of our scheme. Nevertheless, the proposed scheme because of its simplicity and efficiency may be a potential alternative of the widely-used power allocation and SIC strategy.

Acknowledgments. This work was supported in part by the National Natural Science Foundation of China (62201198), in part by the Henan Key R&D and Promotion Projects (232102210156 and 232102210068), and in part by the Key Scientific Research Project Plan of Henan Province Colleges and Universities (23A510016, 23A510018, and 23A520052).

Disclosure of Interests. The authors have no competing interests to declare that are relevant to the content of this article.

References

1. Capez, G.-M., Henn, S., Fraire, J.-A., Garello, R.: Sparse satellite constellation design for global and regional direct-to-satellite IoT services. IEEE Trans. Aerosapce Electron. Syst. **58**(5), 3786–3801 (2022)
2. Guo, C., Chen, X., Yu, J., Xu, Z.: Design of joint device and data detection for massive grant-free random access in LEO satellite internet of things. IEEE Internet Things J. **10**(8), 7090–7099 (2023)
3. Vangelista, L.: Frequency shift chirp modulation: the LoRa modulation. IEEE Signal Process. Lett. **24**(12), 1818–1821 (2017)
4. Ouyang, X., Zhao, J.: Orthogonal chirp division multiplexing. IEEE Trans. Commun. **64**(9), 3946–3957 (2016)
5. Chiani, M., Elzanaty, A.: On the LoRa modulation for IoT: waveform properties and spectral analysis. IEEE Internet Things J. **6**(5), 8463–8470 (2019)
6. Goursaud, C., Gorce, J.-M.: Dedicated networks for IoT: PHY/MAC state of the art and challenges. EAI Endorsed Transactions on Internet of Things 1 (2015)
7. Croce, D., Gucciardo, M., Mangione, S., Santaromita, G., Tinnirello, I.: Impact of LoRa imperfect orthogonality: analysis of link-level performance. IEEE Commun. Lett. **22**(4), 796–799 (2018)
8. Colavolpe, G., Foggi, T., Ricciulli, M., Zanettini, Y., Mediano-Alameda, J.: Reception of LoRa signals from LEO satellites. IEEE Trans. Aerosapce Electron. Syst. **55**(6), 3587–3602 (2019)
9. Ben Temim, M.-A., Ferre, G., Laporte-Fauret, B., Dallet, D., Minger, B., Fuche, L.: An enhanced receiver to decode superposed LoRa-like signals. IEEE Internet Things J. **7**(8), 7419–7431 (2020)
10. Noreen, U., Clavier, L., Bounceur, A.: LoRa-like CSS-based PHY layer, capture effect and serial interference cancellation. In: 24th European Wireless Conference, Catania, Italy, pp. 1–6 (2018)
11. Su, B., Qin, Z., Ni, Q.: Energy efficient uplink transmissions in LoRa networks. IEEE Trans. Commun. **68**(8), 4960–4972 (2020)
12. Nguyen, T.-T., Nguyen, H.-H., Barton, R., Grossetete, P.: Efficient design of chirp spread spectrum modulation for low-power wide-area networks. IEEE Internet Things J. **6**(6), 9503–9515 (2019)
13. Sibbett, T., Moradi, H., Boroujeny, B.: Normalized matched filter for blind interference suppression in filter bank multicarrier spread spectrum systems. IEEE Access **10**, 64270–64282 (2022)
14. Shi, J., Liu, X., Fang, X., Sha, X., Xiang, W., Zhang, Q.: Linear canonical matched filter: theory, design, and applications. IEEE Trans. Signal Process. **66**(24), 6404–6417 (2018)

Remote Sensing Landslide Hazard Risk Analysis Based on Attention Fusion

Tong Li, Liping Liu[✉], Xuehong Sun, Yu Wang, Yiming Jin, Yu Liu, and Feng Qiao

Ningxia University, Yinchuan 750021, Ningxia, China
liuliping8186@126.com

Abstract. As one of the most serious geologic disasters, landslides cause great losses to human lives and properties, so the reliable and timely identification of landslide hazards is of great social and economic significance. In this paper, by adding the attention fusion remote sensing image landslide identification method, using PS-InSAR and SBAS-InSAR two kinds of time series InSAR technology to extract the surface deformation information in the study area, combined with the results of the time series deformation of the algorithm identified in this paper for the risk assessment of landslide potential points, focusing on the analysis of the northern part of the village of Wenbao in Lund County, the western part of the village of Guanglian landslide potential areas, and the use of time series deformation curve quantitatively describe the landslide potential areas. With the help of time series deformation curves, the deformation of the landslide hazardous areas is quantitatively described, which provides a new idea for the identification and risk analysis of landslide hazards.

Keywords: Remote sensing images · Landslide identification · Image recognition · deep learning

1 Introduction

As a serious threat to the safety and sustainable development of human society, geologic disasters, including landslide disasters, have been attracting much attention because of their sudden occurrence and great threat to people's property.

In recent years, with the rapid development of deep learning technology, the introduction of the attention mechanism has brought new ideas and methods for remote sensing data analysis. Through the application of self-attention mechanism, key surface features can be extracted from massive remote sensing data, and potential areas of landslide hazards can be identified more accurately.

The purpose of this paper is to explore the method based on improved attention fusion, combining multi-source remote sensing data and surface information, to realize the comprehensive risk analysis of landslide potential. With the powerful capability of deep learning, we will aim to accurately identify potential landslide hazard areas, analyze their geomorphic features and environmental context, and integrate the spatial and temporal trends to reveal the potential landslide risks in a more comprehensive way. Through this research, we aim to provide a scientific basis for geohazard prevention and control and risk early warning to meet the challenges posed by landslide hazards.

Q. Yu (Ed.): SINC 2023, CCIS 2057, pp. 139–149, 2024.
https://doi.org/10.1007/978-981-97-1568-8_12

2   Landslide Recognition from Remote Sensing Images Based on Attention Fusion

In the self-attention mechanism, the sequence length dominates the self-attention computation, and it is proved theoretically that the feature space matrices Q, K, V formed by the self-attention mechanism are low-rank, so we propose an improved self-attention mechanism that linearly maps the combination of the serialized data K', V' of the original self-attention mechanism to the low-rank matrices K'', $V''^{s \times d}$, where $s \ll n$ and the self-attentive output transformation is:

$$f_{att}(Q', K'', V'') = softmax\left(\frac{Q'K''^{\mathrm{T}}}{\sqrt{d}}\right)V'' \tag{1}$$

The improved self-attention can shorten the sequence length of the original serialized data K', V', so as to achieve the goal of reducing the model parameters and lightweighting the model.

According to the improved self-attention mechanism to design self-attention coding module [1], the structure is shown in Fig. 1. The module inputs semantic feature map $X \in R_{C \times H \times W}$, generates three spatial matrices Q, K, V, and obtains Q', K', V' after convolution operation, and then obtains low-rank matrices K'', V'', after linear mapping, and utilizes Q' to make a query on K' to obtain the query result $Q'K''^T$, and then combines the query with the randomized position matrix is summed and softmax processed to obtain matrix V'' correlation weight between sequence data, in which the random position matrix can be trained by the model to obtain the position-related information between sequence data [2], and finally the correlation weight matrix is multiplied with the value matrix V'' to obtain the self-attention of the semantic feature map. The improved self-attention coding module casn reduce the feature map scale from both channel and space, which reduces the computational cost and realizes the self-attention mechanism inside the feature map.

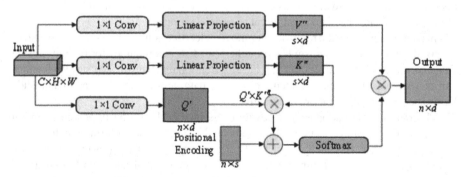

Fig. 1. Improved Self-Attention Coding Module.

The structure of the improved self-attentive decoding module is shown in Fig. 2, which utilizes the low-level semantic feature $X \in R_{C \times H \times W}$ to generate the query input

Q, and utilizes the high-level semantic feature $Y \in R_{C \times h \times w}$ to generate the key and value inputs K and V. After the convolution and linear mapping of the similar coding module to obtain Q', K'', V'', and obtains the correlation weights among sequence data under two semantic scales through the query result $Q'K''^T$ to obtain the correlation matrix between the sequence data under the two semantic scales, add the position information and softmax processing to obtain the correlation weight of feature X reflected on feature Y, and finally multiply the correlation weight with the value V'' matrix to obtain the self-attention feature map [3]. On the basis of inheriting the advantages of the encoding module, the decoding module is able to capture the correlation of different spatial location features of different feature maps and realize the self-attention mechanism among feature maps.

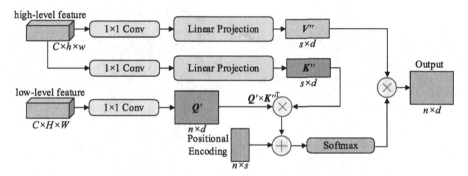

Fig. 2. Improved Self-Attention Decoding Module.

3   InSAR-Based Landslide Hazard Risk Analysis

In this section, permanent scatterer interferometry and short baseline set interferometry are used to extract deformation information in the study area, and the key deformation areas are calibrated and analyzed, and then the time-series InSAR deformation extraction results are combined with this paper's remotely sensed landslide identification algorithm to analyze potential landslide hazards in the study area.

3.1   Definition of Key Deformation Zones in the Study Area

Combining the PS-InSAR and SBAS-InSAR time-series deformation results [4, 5], a total of 185 key deformation areas are circled, as shown in Fig. 3, which correlate the geomorphic features of the study area with the key deformation areas, and analyze the deformation of the study area in terms of geotectonic aspects.

Figure 3 shows that the key deformation zones are mainly distributed in the middle and low mountainous areas, loess hilly areas and red rocky hilly areas, the number of which accounted for 56.22%, 21.62% and 20.00%, respectively, and the development of the key deformation zones in the river valley plain area is less, which accounted for 2.16%, and the statistics are shown in Table 1.

142   T. Li et al.

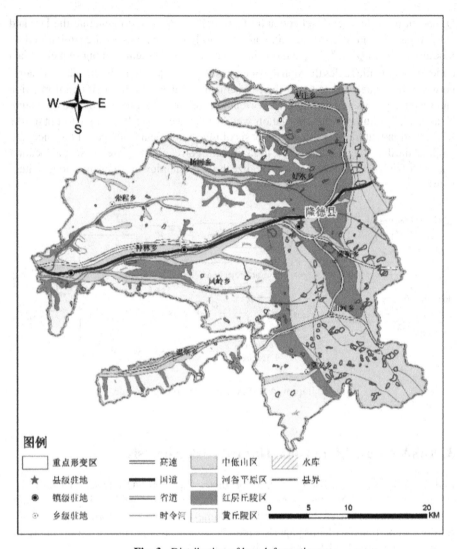

Fig. 3. Distribution of key deformations.

According to Fig. 3 and Table 1, it can be found that the middle and low mountainous areas account for the largest proportion of key deformation areas, which are distributed in the west side of Liupan Mountain, mainly affected by the movement of crustal plates, and the surface deformation is intense. The distribution of key deformation areas in loess hilly areas is the second largest, accounting for 21.62% of the total [6]. Due to the loess has homogeneity and uprightness, under the action of wind and rain, its microfissure extends and expands, greatly reducing the integrity of loess, coupled with the action of gravity, by the tensile stress and shear stress, along the existing fissures, vertical joints, tectonic joints or loess layer and the weak interface between the ancient soil constantly

Table 1. Statistics of geomorphological features of deformation zones.

Serial number	Landform type	Hazard class for key deformation zones	Number of priority deformation zones	Subtotal	Percentage
1	River valley plains region	Low	2	4	2.16%
		High	2		
2	Red layer hilly area	Low	10	37	20.00%
		Middle	18		
		High	9		
3	Hilly area	Low	17	40	21.62%
		Middle	22		
		High	1		
4	Mid-to-low mountain area	Low	47	104	56.22%
		Middle	7		
		High	50		
	Total	-	185	185	100%

produce slip, pull off and tracking shear, gradually forming a sliding surface, under the action of earthquakes and heavy rainfall Under the action of earthquakes and strong rainfall, landslides and avalanches eventually occur. The key deformation area in the red rocky hills area accounts for 20.00% of the total. The topography of this area is highly undulating, and the weathering of the mudstone is serious, so landslides, avalanches, and unstable slopes (landslide hazards and avalanche hazards) are found in this area [7]. The distribution of key deformation zones in the river valley plain area is smaller, accounting for only 2.16% of the total number of disasters, which is mainly due to the fact that its terrain is flat and open, and thus the geo-environmental conditions for the occurrence of landslides, avalanches and other disasters are weaker.

3.2 Landslide Risk Analysis

The surface deformation inversion results from PS-InSAR and SBAS-InSAR show that the western foothills of Liupan Mountain and the southern part of Lund County in Lund County show a subsidence trend along the radar line-of-sight direction, while the northwestern part of Lund County, which is far away from the Liupan Mountain, mainly shows an uplift trend. Taking the deformation areas in the northern part of Wenbao Village, the western part of Guanglian Village, and the northern part of Xujiazhuang in Lund County as examples, we analyze whether there are landslide hazards in these areas by combining the deformation extracted from the time-series InSAR.

The deformation zone in the northern part of Wenbao Village is located at 35°27′38.66″N, 105°56′48.86″E, with an elevation of 2013 m, and the deformation zone has a northeast-southwest orientation, and the annual average deformation rate of this area is 22 mm/y according to the deformation information of SBAS-InSAR. The deformation area in the northern part of Wenbao village is shown in Fig. 4(a), and Fig. 4 (b) corresponds to the potential landslide hazard area detected by the recognition algorithm in this paper, from which it can be seen that the predicted area of landslide hazard is smaller compared to the deformation area, and it is mainly concentrated in the part of the deformation area that is at a higher altitude, which is mainly due to the fact that when slow landsliding occurs, the deformation is the first to occur in the high altitude of the landslide body, and thus the topographic characteristics of the landslide hazard are consistent with the model's estimation of landslide targets [8]. Table 2 shows the length-area correspondence between the landslide hazard area and the deformation area, where the landslide hazard area accounts for 24.5% of the deformation area.

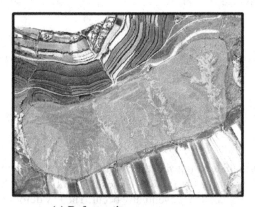

(a) Deformation areas

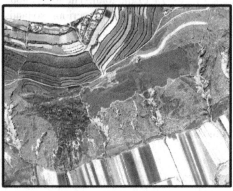

(b) Landslide hazards

Fig. 4. Deformation Area North of Winnebago Village

From the SBAS-InSAR time-series deformation results, Fig. 4 shows the time-series deformation curves of the landslide hazard area. The horizontal axis represents the date time, ranging from January 2019 to May 2021, corresponding to the data acquisition

Table 2. Parameters of deformation zone in the northern part of Winborough Village.

Area	Maximum north-south length/m	Maximum east-west length/m	Area covered/m^2
Landslide hazard	133.3	433.6	39680.9
Region of deformation	269.7	733.5	162109.8

period for deformation extraction. The vertical axis represents the deformation variables in mm/12d, as the revisit period of Sentinel I data is 12 days, and the deformation variables of neighboring revisits are accumulated together. The figure shows an overall sinking trend, but an uplifting phenomenon from December 2019 to May of the following year. This phenomenon is mainly caused by two factors: first, on December 13, 2019 Lund County received snowfall, and the C-band microwaves had difficulty penetrating the snow layer, resulting in temporary topographic uplift; second, the temperature in Lund County continued to be sub-zero in January 2020, and the moisture condensed into frost, which also resulted in temporary topographic uplift. As the temperature rises, vegetation growth affects after March, and the terrain lifts again briefly [10]. After the vegetation growth saturated, around August, the area again appeared to subside. In summary, the landslide hazard area in the northern part of the village of Winborough has always been in a subsidence trend, and the area should be given priority attention (Fig. 5).

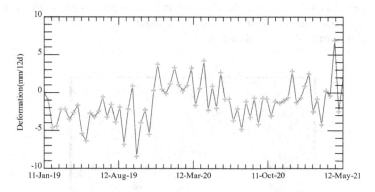

Fig. 5. Temporal deformation curve of the landslide hazard area in the northern part of Wenbao village.

The deformation zone in the western part of Guanglian Village is located at 35°36'22.92"N, 106°1'6.66"E, with a sea wave of 2034 m, and the deformation zone generally shows a northeast-southwest direction, and the annual average deformation rate of the area is 36 mm/y as shown by the results of the SBAS-InSAR deformation.

The deformation area in the western part of Guanglian Village is shown in Fig. 6(a), and Fig. 6 (b) corresponds to the potential landslide hazard area of the recognition algorithm. It can be seen that the landslide prediction area is more dispersed compared

(a) Deformation areas   (b) Landslide hazards

Fig. 6. Western Deformation Zone of Kwang Luen Village.

to the deformation area, and it is mainly distributed in the slope area with obvious optical features, which is caused by two reasons, namely, the optical and elevation aspects. First of all, the western part of Guanglian Village is a complex hilly structure, which is similar to the edge structure of the landslide target in terms of optical features; from the perspective of elevation, the deformation area has experienced a long subsidence process, and the terrain tends to be similar to that of the landslide target in the shape of a tongue or a funnel, so to some extent, part of the subsidence area in the deformation area can be regarded as a slow-change landslide target [11]. In order to qualitatively analyze the landslide hazard identification area, the above hazard area is divided into seven parts, and the location of each part is shown in Fig. 7.

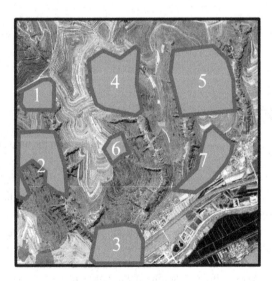

Fig. 7. Distribution of landslide hazards in the western part of Guanglian Village.

Table 3 shows the length and area information of deformation areas and landslide hazards. By analyzing Fig. 7 and Table 3, it can be found that the area of the No. 4 and No. 5 hazards accounts for the largest proportion of the total area of the landslide hazards, reaching 48.3% of the total area of the landslide hazards, and the length of landslide hazards is in the order of one hundred meters, which reflects that this paper's landslide recognition algorithm is able to extract the landslide hazards of the order of square kilometers.

Table 3 shows the time-sequence deformation curves of the seven landslide potential points in the region. From the time-varying curves, it can be seen that the trend of the deformation curves of Point 3 and Point 7 near Guanglian Village are basically the same, and they all briefly lifted up during the period of February to April, 2019, which is mainly related to the continuous snowfall situation in that period.

Table 3. Parameters of deformation area in the west of Guanglian village.

Area	Maximum north-south length/m	Maximum east-west length/m	Area covered/m^2
Landslide Hazard #1	392.2	410.8	126369.5
Landslide Hazard #2	736.4	535.0	243256.1
Landslide Hazard #3	535.1	692.7	258604.4
Landslide Hazard #4	778.8	612.4	354414.2
Landslide Hazard #5	769.1	764.5	512941.0
Landslide Hazard #6	325.0	291.6	65421.4
Landslide Hazard #7	726.2	668.8	233716.2
Region of deformation	2627.4	2417.6	3779366.1

It can be seen that the seven potential points have been lifted up to varying degrees in this period of time, and each potential point Point 2 is located in the high-elevation slope area, and the time-varying curve shows that the deformation characteristics of this area are obvious, and it shows an accelerated subsidence trend, which should be listed as a key area of concern. Points 1, 4 and 6 showed a sudden elevation during April and May 2020, which is due to the existence of continuous rainfall in Lunde County during the same period, and in May 2020, the number of days of rainfall in Lunde County reached 1,000 days, and the number of days of rainfall in Lunde County reached 1,000 days. The number of rainfall days in Lund County reached 14 days, which greatly accelerated the growth process of the local vegetation, which in turn affected the temporal curves of these hidden points. As a whole, the landslide hazard sites in this deformation zone show a subsidence trend, and measures need to be taken to prevent landslide risks in this area (Fig. 8).

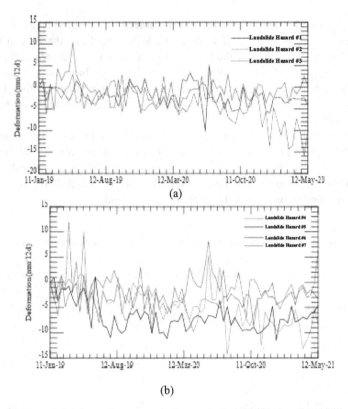

Fig. 8. Temporal deformation curves in the western part of Kwang Luen Village.

4 Conclusion and Future Work

Fusing InSAR and deep learning methods, not only realized the accurate identification of landslide hazardous areas, but also used the deformation information of the study area to analyze the temporal deformation of landslide hazards, and the accuracy of landslide identification reached 96.81%, and the average accuracy of pixel segmentation reached 90.11%, which compared with the DeeplabV3 + and U-net methods in terms of mIoU and mPA has been improvement, proving the effectiveness of the attention fusion method in landslide identification. Firstly, combining the optical remote sensing images and DEM elevation data of the study area, the location information of landslide hazards in the study area is successfully extracted by using the remote sensing landslide recognition algorithm proposed in this paper; subsequently, two time-series InSAR techniques, PS-InSAR and SBAS-InSAR, are used to obtain the surface deformation information in the study area; finally, taking the northern part of Wenbao Village, the western and southern part of Guanglian Village as an example of the deformation area, the effectiveness of the The feasibility of the combination of deep learning landslide identification method and InSAR technique in landslide hazard identification and analysis.

Acknowledgments. This study was funded by 《A study on vertical structure modeling and inversion method for satellite-borne L/P dual-band Po-InSAR vegetation》 (grant number 62061038).

References

1. Guo, C., Xu, Q., Dong, X.: Landslide identification based on SVF terrain visualization method--a case study of typical landslide in Danba County, Sichuan Province. J. Chengdu Univ. Technol. (Nat. Sci. Edn.) **48**(06), 705–713 (2021)
2. Jian, X., Zhao, K., Zuo, X.: Landslide hazard recognition based on faster R-CNN target detection - a case study of urban area in Fugong County, pp. 1–9. Chemical Minerals and Processing (2022)
3. Wang, Y., Zhang, P., Sun, K.: Remote sensing landslide target recognition based on attention fusion. Liq. Cryst. Disp. **37**(11), 1498–1506 (2022)
4. Hu, F., Fan, Y.: Application research on landslide disaster identification by remote sensing images. Agric. Disaster Res. **11**(04), 98–99 (2021)
5. Lu, H., Li, W., Xu, Q., et al.: Early identification of landslide hazards upstream and downstream of Jinshajiang Baige landslide by combining optical remote sensing and InSAR. J. Wuhan Univ. (Inf. Sci. Edn.) **44**(09), 1342–1354 (2019)
6. Wu, X.: Research on Object-Oriented Convolutional Neural Network Landsat8 Image Classification Method. Wuhan University, Wuhan (2019)
7. Wang, W.H.: Landslide hazard identification based on SBAS-InSAR and machine learning—taking Lanzhou City as an example. Lanzhou Jiaotong University, Lanzhou (2021)
8. Zhang, Y.L., Fu, Y.H., Sun, Y.: Landslide detection from high-resolution remote sensing image using deep neural network. Highway **66**(5), 188–194 (2021)
9. Zhang, P.F.: Research on automatic extraction of remote sensing images of co-seismic landslides. Institute of Geology, China Earthquake Administration, Beijing (2021)
10. Liao, M.S., Dong, J., Li, M.H.: Radar remote sensing for potential landslides detection and deformation monitoring. Nat. Remote Sens. Bull. **25**(1), 332–341 (2021)
11. Xu, B.Z., Zhu, Q., Li, H.F.: A landslide extraction method of remote sensing image based on multi-scale depth attention model. J. Geomatics **47**(3), 108–112 (2022)

A Multi-UAVs Cooperative Spectrum Sensing Method Based on Improved IDW Algorithm

Jie Shi[1], Jingzheng Chong[1], Zejiang Huang[1], and Zhihua Yang[1,2][(✉)]

[1] Harbin Institute of Technology(Shenzhen), Shenzhen 518055, China
`yangzhihua@hit.edu.cn`
[2] PengCheng Laboratory, Shenzhen 518055, China

Abstract. With the development of spatial information networks, the perceptual allocation of spectrum resources has become a research hotspot. To solve the problems of low accuracy and efficiency in traditional single uav spectrum sensing, a multi-UAVs collaborative spectrum sensing method based on improved IDW algorithm is proposed in this paper. Firstly, a spectrum sensing model for collaborative exploration of multiple UAVs was constructed; Secondly, a spectral intensity cost factor is added to the cost function of UAV path planning, enabling UAVs to explore electromagnetic environment more efficiently; Finally, the accuracy of spectrum data completion is improved by combining IDW algorithm with propagation model. The simulation results show that the task completion time is reduced compared to current advanced path planning methods, and the accuracy of the completion algorithm is nearly 10dB higher than that of IDW and tensor completion methods, which has high practical value.

Keywords: Unmanned Aerial Vehicle · Spectrum Sensing · Path Planning · Inverse Distance Weighting Method

1 Introduction

The spatial information network is a network system that uses the space platform as a carrier to acquire, transmit and process spatial information in real time. With the development of spatial information networks, the spectrum environment has become increasingly complex, resulting in the scarcity of electromagnetic spectrum resources and serious spectrum security conditions, which has brought new challenges to spectrum management. The issue of how to effectively utilize limited spectrum resources and combat and prevent illegal electromagnetic equipment has received considerable attention.

Spectrum Sensing provides a way to solve the above problems. Based on spectrum data, it understands the current status of the electromagnetic environment in a specific area and predicts future trends. The electromagnetic spectrum map is a visualization method that accurately reflects the distribution of the electromagnetic spectrum in a specific space from multiple angles [1]. Spectrum Sensing

Q. Yu (Ed.): SINC 2023, CCIS 2057, pp. 150–163, 2024.
https://doi.org/10.1007/978-981-97-1568-8_13

generally achieves the purpose of obtaining the spectrum situation by mapping the distribution of radio parameters such as received signal strength (RSS) in a specific space. The construction of spectrum maps is of great significance for the allocation of spectrum resources and the comprehensive management of the electromagnetic environment.

In this paper, we propose a multi-UAVs collaborative spectrum sensing method in an unknown environment based on an improved IDW algorithm. The measurement data of key locations in the area are obtained through UAV formation, and the improved IDW algorithm is used to estimate the missing data and construct a highly accurate Full spectrum map of degrees. The main contributions of this article are as follows:

(a) A multi-UAVs cooperative spectrum sensing scheme is proposed. Utilize multi-UAV path planning to efficiently obtain spectrum data, and adjust the UAV trajectory in real time through measurement data to obtain accurate values of signal strength near the radiation source. This is consistent with the characteristics of the interpolation algorithm and facilitates more accurate construction of spectrum maps.
(b) An improved IDW algorithm is proposed for completion of spectrum data. The classic IDW algorithm is improved based on measurement data, channel models and angle factors to estimate unknown spectrum data with higher accuracy. The constructed spectrum map has significant performance improvements compared to traditional methods.

2   Related Works

To construct a spectrum map, spectrum measurement equipment to collect raw data is needed. In most existing methods, the original data collection method is to deploy a large number of spectrum sensors in a specific area. This type of measurement method requires a large amount of sensor deployment, and the sensors are not mobile. However, this method can only collect spectrum data within a fixed range. If the collection area changes, the sensors need to be re-collected and deployed. UAVs can be used for data collection and measurement due to their high mobility. As the area to be measured increases, multi-UAV collaborative path planning algorithms are proposed in different scenarios. A UAV path planning method based on artificial potential fields is proposed in [2,3]. A UAV path planning method based on voxel maps [4,5], which allows UAVs to evenly distribute exploration areas and avoid collisions between UAVs. A method based on genetics [6] enables the drone to reduce flight energy consumption as much as possible. However, there are currently few path planning methods designed for better spectrum sensing.

Since the raw data collected by drones are spatially discrete, it is necessary to estimate the vacancy value based on the raw data to complete the entire spectrum map. Spectrum map completion methods are usually divided into two categories, namely direct methods and indirect methods. The direct method performs map completion based on original data, among which the most common method is Inverse Distance Weight (IDW) [7]. Melvasalo, M and Mao, D. et al. used Kriging spatial interpolation method to predict missing data in [8,9]. Tang, M. et al. proposed a tensor completion method combined with a priori model [10], which can obtain the complete spectrum of the measurement area. Huang, X. and colleagues introduced an innovative spectrum mapping strategy designed for extensive cognitive radio networks (CRNs) [11]. This approach leverages past outcomes of spectrum decisions to enhance its effectiveness. Unlike direct methods that obtain missing data from raw data, indirect methods are driven by channel models. Methods based on Location Estimate (LE) [12] and signal-to-noise ratio-assisted methods [13] are proposed. Both of these algorithms employ pre-existing data about the electromagnetic environment, including radio propagation models, to enhance the efficiency of data recovery. Sato, K. and co-authors utilize Kriging spatial interpolation for real-time estimation of shadow fading [14]. Additionally, Isselmou, Y.O., and colleagues initially derived the initial spectrum map using prior information from the measurement area and subsequently refined it using actual measurement data [15]. In addition to the two methods mentioned above, neural networks are applied to the construction of electromagnetic spectrum situations. Song Wenjia used Long Short-Term Memory (LSTM) neural networks to perceive and judge UAV communication spectrum [16]. Zhang Han completed the construction of a specific electromagnetic environment spectrum map by establishing and training a Residual Autoencoder (RA) [17].

Many direct methods rely on spatial interpolation algorithms. Due to the nature of these algorithms, the estimated value of an unknown element often falls between the maximum and minimum values of the measured data. However, when the known data is concentrated in a specific part of the actual area, all estimated values must be within the range of this portion, inevitably leading to a decrease in algorithm performance. Moreover, indirect methods like the mentioned Cognitive Radio Networks (CRNs) generally exhibit superior performance compared to direct methods, and the algorithm's effectiveness remains unaffected by whether the spectrum data collected is evenly distributed across the measurement area. The indirect method relies heavily on the prior information of the electromagnetic environment, and cannot achieve satisfactory results when constructing spectrum maps in unknown areas that lack prior information. Moreover, among the indirect methods introduced in most existing literature, these approaches operate under the assumption of a single radiation source throughout the entire measurement area, a premise that may not align with real-world scenarios. Using neural networks to construct spectrum maps does not require modeling of the electromagnetic propagation environment or the use of spatial interpolation algorithms. The network model can obtain better results after being fully trained. However, neural network systems such as RA require sufficient existing data for training, which requires a large amount of complete spectrum data in different electromagnetic propagation environments in actual scenarios, which is not feasible and still needs to be further optimized. In addition, most of the current literature on spectrum map construction focuses on the spectrum map completion stage after obtaining measurement data, often without considering the work of the data collection stage, and failing to adjust the measurement method according to the characteristics of the data processing algorithm used. And optimization, taking into account the direct method's high requirements for data distribution uniformity, a specific method needs to be used in the measurement stage to make the measurement data cover the largest possible range.

3   System Model

The multi-UAVs collaborative spectrum mapping system proposed in this article is shown in Fig. 1. Under the premise that the radiation source is unknown, multiple UAVs are located at a fixed height above the area to be measured, carrying out completely autonomous environmental exploration and path planning, and continuously collecting the RSS values of the current location through a spectrum analyzer mounted on the drone. The two-dimensional coordinates at this height are marked as (x_i, y_i), where $i \in [1, 2, \cdots, K]$ is the UAV number, and x_i, y_i are respectively the horizontal and vertical axis coordinates of the UAV in the two-dimensional space.

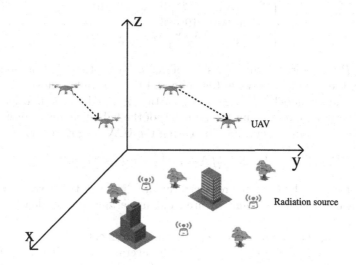

Fig. 1. Multi-UAVs collaborative spectrum mapping system diagram.

In order to enable multiple UAVs to complete exploration tasks efficiently and safely, this paper uses UAV flight time and flight distance as optimization indicators. First consider the optimization of UAV flight time. A key issue in UAV path planning is to ensure that the UAV completes the exploration mission within a limited time. Therefore, the time from the UAV starting to explore to the last UAV ending the exploration should be as short as possible, which reduces the energy consumption of the UAV needs to make the flight distance of the drone as short as possible.

$$\begin{cases} J_T = \min[\max(t_i)] \\ J_D = \min \sum_{i=1}^{K} d_i \end{cases} \tag{1}$$

in which t_i is the time for the i-th UAV to complete the exploration, and d_i is the flight distance of the i-th UAV.

The spectrum map in this paper uses the form of a heat map to display the received RSS distribution. In order to collect and process data more conveniently, this article constructs a three-dimensional rectangular coordinate system based on the starting point and end point of the area to be measured. This coordinate system divides the area to be measured into $N_1 \times N_2 \times N_3$ cubes, then the coordinates of the cube with serial number (n_1, n_2, n_3) can be expressed as $((n_1 - 0.5) \times d_1, (n_2 - 0.5) \times d_2, (n_3 - 0.5) \times d_3)$, where d_1, d_2 and d_3 are respectively the length, width, and height of the set cube, and each cube stores the RSS of the position value.

RSS is contingent upon both the propagation channel model and the transmit power of all radiation sources present within the measurement area. The ideal receiving spectrum intensity of a certain cube can be declared as:

$$\begin{cases} P_i^{rx}[\text{dBm}] = P_i^{tx}[\text{dBm}] - L_i[\text{dB}] \\ P_i^{rx}[\text{mW}] = 10^{P_i^{rx}[\text{dBm}]/10} \\ P_i^{rx}[\text{mW}] = \sum_{i=1}^{N_{tx}} P_i^{rx}[\text{mW}] \end{cases} \tag{2}$$

in which P_i^{tx} represents the emission power of the i-th radiation source, L_i represents the path loss between the cube and the i-th radiation source.

The path loss model (PL) used in this article is based on the free space path loss model, taking into account the influence of the fading channel caused by the normal shadow environment, and increasing the UAV height factor.

$$PL[dB] = 32.4 + 20 lg(f_c) + 10(I_A + h_{UAV}^{I_B}) \cdot lg(d) + \chi_\sigma \tag{3}$$

in which d , f_c and h_{UAV} are respectively the distance, the carrier frequency and height of the UAV. I_A and I_B are random parameters decided by the circumstance.

Considering the possibility of repeated measurements of the same cube by multiple drones, it is reasonable to take the average of multiple measurement results as the final measurement value. Therefore, the final RSS measurement of the cube can be expressed as

$$P^{mea} = \frac{1}{N_{mea}} \sum_{k=1}^{N_{mea}} P^{meak} \tag{4}$$

in which P^{mea} and P^{meak} represent the final measurement value and the k-th measurement value of the cube RSS respectively, N_{mea} is the number of measurements of the cube.

The RSS value of the unknown cube is estimated using an interpolation algorithm. The estimated value depends on the measurement value of the known cube and its corresponding weight value. The unknown cube RSS estimate $\hat{P}$ can be expressed as

$$\hat{P} = \sum_{j=1}^{N} \omega_j P_j^{\text{mea}} \tag{5}$$

in which ω_j represents the weight of the j-th measurement value to the estimated value, P_j^{mea} represents the j-th RSS measurement value, and N represents the number of measurement values.

Through estimation and calculation, the RSS values of all unknown cubes can be obtained, and a complete spectrum intensity matrix can be constructed. In order to reflect the accuracy of the constructed spectrum map, the root mean square error (RSE) can be used to represent it as

$$RSE(\text{dB}) = 10 \log_{10} \frac{\left\| \tilde{M} - M \right\|_2}{\|M\|_2} \tag{6}$$

in which M is theoretical spectral intensity matrix, $\tilde{M}$ is the recovered spectral intensity matrix.

The overall flow chart of spectrum map construction in this paper is shown in Fig. 2. First, multiple UAVs autonomously explore and exchange information in real time within a set unknown area, and uniformly collect spectrum intensity data according to the planned path. Then based on the collected spectrum intensity data, data is completed through the spectrum completion algorithm. Finally, generating a completed electromagnetic spectrum map.

4 Algorithm Description

4.1 Multi-UAVs Collaborative Path Planning Method Based on Spectrum Intensity

In order to improve the exploration efficiency of the system, this paper proposes a multi-UAVs collaborative path planning method based on spectrum intensity. First, updating the map boundary based on the status of the UAV and sensor data; secondly, considering the flight level, boundary level, and the changing trend of spectrum intensity to conduct global path planning comprehensively; finally, generating high-quality local path results for UAVs based on the global path planning results. The results of path planning will be submitted to the flight control for execution, and the above process will continue according to the re-planning strategy.

In this paper, a boundary-based method is employed for path planning, where the rationality of the boundary selection sequence plays a crucial role in determining the efficiency of the entire exploration process. Numerous approaches

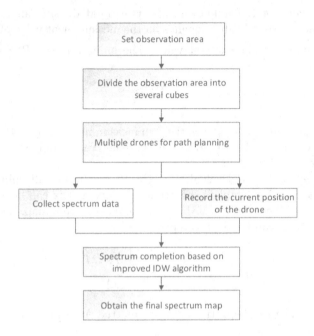

Fig. 2. Multi-UAVs collaborative spectrum mapping system diagram.

utilize the Traveling Salesman Problem (TSP) to establish the boundary selection sequence. However, the majority of methods only consider the Euclidean distance between boundaries as the cost of TSP, which, although simple, is often insufficient. Moreover, the traditional TSP method stipulates that multiple drones should start from the same starting point, which has limitations. In the application scenario of this article, another asymmetric TSP method is used so that the drone can start from different starting points, and in order to obtain the spectrum intensity more efficiently, the drone should explore in the direction where the spectrum intensity becomes larger to reduce the range is explored to better recover the spectral map and determine the approximate location of the radiation source. Therefore, this paper adds the spectrum intensity cost factor to the boundary cost function of UAV path planning to generate a better global exploration path by solving the asymmetric TSP problem.

As shown in Fig. 3, the drone flies from the previous position p_{-1} to the current position p_0. The spectrum intensity collected at the previous location is RSS_{-1}, and the spectrum intensity currently collected is RSS_0. From this, the spectrum intensity difference between the current location and the previous location can be calculated as

$$\Delta RSS = RSS_0 - RSS_{-1} \tag{7}$$

If ΔRSS is positive, the flight direction of the UAV at the last moment is the direction in which the spectrum intensity increases. Otherwise, it is the direction in which the spectrum intensity decreases. As the target viewpoint of the UAV at the next moment V_i, in order to make the UAV explore in the direction where the spectrum intensity becomes larger, this paper calculates the spectrum intensity cost factor corresponding to each boundary viewpoint as

$$c_b(k) = \begin{cases} -1 & \Delta RSS \cdot \cos\theta_{0k} < 0 \\ 1 & \Delta RSS \cdot \cos\theta_{0k} > 0 \end{cases} \tag{8}$$

in which θ_{0k} is the angle between the lines connecting three points p_0, p_{-1} and V_i.

Finally, this paper will integrate $c_b(k)$ and the flight level factors used in the literature [18] into the cost matrix of the ATSP problem as

$$M_{\mathrm{tsp}}(0, k) = t_{\mathrm{lb}}(V_0, V_k) + \omega_c \cdot c_b(k) \tag{9}$$

in which $t_{\mathrm{lb}}(V_0, V_k)$ is the flight time factor between the two viewpoints and ω_c is the weight assigned to $c_b(k)$.

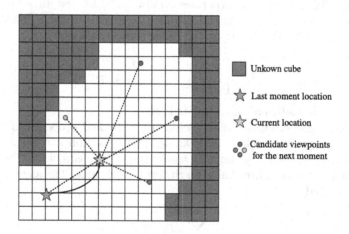

Fig. 3. Viewpoint selection diagram.

4.2 Construction of Spectrum Map Based on Improved IDW Algorithm

UAVs are used to collect the RSS values of some sampling points in a specific area, and based on the known data, the RSS values of unknown points are estimated using the weighted interpolation method. The estimated RSS value can be expressed as

$$\hat{P}_{loc_0} = \sum_{i=1}^{N} \omega_i P_{loc_i} \tag{10}$$

in which loc_0 represents the location needed to be estimated, loc_i represents the i-th known measurement point, $\hat{P}_{loc0}$ is the estimated RSS, P_{loc_i} represents the RSS in loc_i.

The weight value of the traditional IDW algorithm can be expressed as

$$\omega_i = \frac{\frac{1}{d_i^n}}{\sum_{i=1}^{N} \frac{1}{d_i^n}} \tag{11}$$

in which d_i is represents the distance between loc_0 and loc_i, n denotes the power parameter.

The traditional IDW method is based on the assumption that unknown points are close to known measurement points, but the reality of constructing spectrum maps does not always meet this condition. At the same time, this method only considers the influence of distance and ignores other factors that affect RSS in a real electromagnetic propagation environment. Due to these defects, the completion of spectrum maps using the traditional IDW algorithm does not work well. In response to the above problems, this paper improves the weight calculation method of the IDW algorithm from the following aspects.

Initially, the algorithm incorporates an optimized inverse distance weighted interpolation method known as the modified Shepard's method (MSM), which is based on mathematical functions [7]. This is expressed as:

$$p_i = \begin{cases} \frac{1}{d_i} & 0 < d_i \leq \frac{R}{3} \\ \frac{27}{4R}(\frac{d_i}{R}-1)^2 & \frac{R}{3} < d_i \leq R \end{cases} \tag{12}$$

in which R is the radius of the estimated area.

Secondly, considering the angle factor in the method proposed in [19]. This method calculates the angle of each known point position and each estimated point position, and uses the angle factor as one of the parameters that control the influence of the measured value on the estimated value. The angle coefficient can be expressed as

$$angle_i = \frac{\sum_{j=1,j\neq i}^{N} P_i \cdot (1 - \cos\phi_{ij})}{\sum_{j=1,j\neq i}^{N} P_i} \tag{13}$$

in which ϕ_{ij} is the angel between loc_0 and loc_i.

Finally, considering the propagation environment characteristics of the measured area, the channel propagation model factor is included as one of the factors in calculating the weight value as

$$l_i = \frac{L_i^{-1}}{\sum_{j=1}^{N} L_j^{-1}} \tag{14}$$

in which l_i is the path loss between loc_0 and loc_i, determined by the channel propagation model used in (3).

Considering all the above factors, the weight of the RSS value to the predicted value is

$$\omega_i = \frac{P_i^{w_1} \cdot (1+a_i)^{w_2} \cdot (1+l_i)^{w_3}}{\displaystyle\sum_{j=1}^{N} p_i^{w_1} \cdot (1+a_i)^{w_2} \cdot (1+l_i)^{w_3}} \tag{15}$$

in which w_1, w_2 and w_3 are the distance factor, angel factor and path loss factor respectively.

Substituting the weight value w_i into equation (10), the estimated value of the unmeasured point $\hat{P}_{loc_0}$ can be calculated.

5 Simulation Results

Generally speaking, on-site spectrum collection by UAV is expensive and has high errors in actual scenes, so a modeling method based on ray tracing (RT) simulation is considered as an alternative. For the spectrum completion algorithm proposed in this article, the Matlab platform was used for simulation verification. The scene settings were as follows: Harbin Institute of Technology Shenzhen Campus was selected as the experimental scene, and multiple radiation sources were randomly set up in the scene. The area of the reconstructed spectrum map is set to a square, and divided into grids. Each grid is a square with a side length of 1m. The spectrum of the signal source is set to 20MHz, and the power of the radiation source is 0dB. Through theoretical analysis, The spectrum map in this scenario is shown in Fig. 4. The number of UAVs is set to 3, the safe distance for drone flight is set to 10m, and the flight parameters of each drone are the same. When flying in a straight line, the drone flies at a constant speed, the speed is set to 5m/s, and the flight height is always 50m.

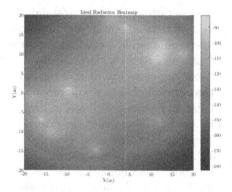

Fig. 4. Viewpoint selection diagram.

According to the path planning algorithm proposed in this article, the UAV autonomously plans the path according to changes in spectrum intensity. We compare the algorithm proposed in this article with three of the most advanced path planning algorithms, including FUEL, Aeplanner, and NBVP. This article uses the open source programs and default parameter settings of these three algorithms to simulate in the scenario we set. The simulation results are shown in Table 1.

Table 1. Path planning simulation results.

path planning algorithm	exploration time(s)				flight distance(m)			
	Avg	Std	Max	Min	Avg	Std	Max	Min
Aeplanner	418	49	422	203	202	26	255	142
NVBP	680	143	941	426	305	62	422	205
FUEL	154	11	188	148	209	16	246	203
Proposed	125	9	157	103	156	11	180	135

As can be seen from Table 1, the algorithm proposed in this paper has significantly shortened the exploration time and flight distance. The main reason is that the algorithm proposed in this article allows the UAV to explore areas with greater spectrum intensity based on the spectrum intensity, and zoom out in real time. This proves that the UAV path planning algorithm proposed in this article is suitable for spectrum surveying and mapping application scenarios.

The collected spectrum data is shown in Fig. 5(a). In order to verify the effectiveness of the spectrum completion algorithm proposed in this article, the final completion result is obtained through the spectrum completion algorithm based on the spectrum data collected by the UAV. As shown in Fig. 5(b), in order to calculate the performance of spectrum map reconstruction, its RSE is calculated.

In order to compare the performance of spectral data processing methods, a large proportion of points need to be removed from the spectrum map. Therefore, We sample the original spectrum map at a random sampling rate. This sampling rate represents the percentage of RSS that are remained in the spectrum map. In the simulation, this paper compares the RSE of the improved IDW algorithm with the classic IDW algorithm and tensor completion algorithm at different sampling rates. The simulation results are shown in Fig. 6.

Through comparison, it can be found that the error of spectrum reconstruction decreases as the sampling rate increases. Under different sampling rates, the method proposed in this article is more accurate and stable than both IDW and tensor completion methods. This demonstrates that the enhanced IDW algorithm proposed in this article has significantly increased the accuracy of spectrum reconstruction by nearly 10dB.

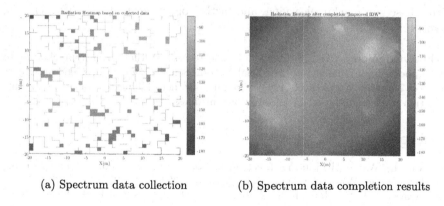

(a) Spectrum data collection   (b) Spectrum data completion results

Fig. 5. Spectrum recovery renderings

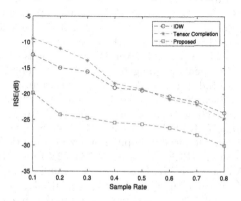

Fig. 6. Viewpoint selection diagram.

6   Conclusion

In view of the problems of low accuracy and low efficiency in spectrum sensing by a single UAV, a multi-UAVs collaborative spectrum sensing method based on the improved IDW algorithm is proposed. UAVs carrying spectrum analyzers start from different starting points at the same height according to the cost. The function explores the unknown area, collects spectrum data, uses the improved IDW algorithm combined with the propagation model to complete the spectrum data, and finally restores the spectrum map of the unknown area. Through simulation, it can be found that the multi-UAVs path planning algorithm proposed in this article can effectively improve the efficiency of UAV exploration, and the improved IDW algorithm is significantly improved compared to the traditional IDW and tensor completion algorithm. This article only considers the spectrum intensity at a single frequency. Subsequent research will further study scenarios with more complex spectrum environments, explore a spectrum recovery algorithm based on machine learning, and further optimize the UAV path planning algorithm.

Acknowledgments. The authors would like to express their high appreciations to the supports from Guangdong Provincial Natural Science Foundation General Project (Project Number: 2021A1515011953): Satellite Internet of Things data cache and distribution mechanism based on information timeliness

References

1. Guo, L., Wang, M., Lin, Y.: Electromagnetic environment portrait based on big data mining. Wirel. Commun. Mob. Comput. **2021**, 1–13 (2021)
2. Jayaweera, H.M., Hanoun, S.: A dynamic artificial potential field (d-APF) UAV path planning technique for following ground moving targets. IEEE Access **8**, 192760–192776 (2020)
3. Sulieman, M.H., Gursoy, M.C., Kong, F.: Antenna pattern aware UAV trajectory planning using artificial potential field. In: 2021 IEEE/AIAA 40th Digital Avionics Systems Conference (DASC), pp. 1–7. IEEE (2021)
4. Zhao, Y., Yan, L., Xie, H., Dai, J., Wei, P.: Autonomous exploration method for fast unknown environment mapping by using UAV equipped with limited FOV sensor. IEEE Trans. Ind. Electron. **71**(5), 4933–4943 (2023)
5. Zhou, B., Xu, H., Shen, S.: Racer: rapid collaborative exploration with a decentralized multi-UAV system. IEEE Trans. Robot. **39**, 1816–1835 (2023)
6. Asim, M., Mashwani, W.K., Belhaouari, S.B., Hassan, S.: A novel genetic trajectory planning algorithm with variable population size for multi-UAV-assisted mobile edge computing system. IEEE Access **9**, 125569–125579 (2021)
7. Denkovski, D., Atanasovski, V., Gavrilovska, L., Riihijärvi, J., Mähönen, P.: Reliability of a radio environment map: Case of spatial interpolation techniques. In: 2012 7th International ICST Conference on Cognitive Radio Oriented Wireless Networks and Communications (CROWNCOM), pp. 248–253. IEEE (2012)
8. Melvasalo, M., Koivunen, V., Lundn, J.: Spectrum maps for cognition and co-existence of communication and radar systems. In: 2016 50th Asilomar Conference on Signals, Systems and Computers, pp. 58–63. IEEE (2016)
9. Mao, D., Shao, W., Qian, Z., Xue, H., Lu, X., Wu, H.: Constructing accurate radio environment maps with kriging interpolation in cognitive radio networks. In: 2018 Cross Strait Quad-Regional Radio Science and Wireless Technology Conference (CSQRWC), pp. 1–3. IEEE (2018)
10. Tang, M., Ding, G., Wu, Q., Xue, Z., Tsiftsis, T.A.: A joint tensor completion and prediction scheme for multi-dimensional spectrum map construction. IEEE Access **4**, 8044–8052 (2016)
11. Huang, X.L., Gao, Y., Tang, X.W., Wang, S.B.: Spectrum mapping in large-scale cognitive radio networks with historical spectrum decision results learning. IEEE Access **6**, 21350–21358 (2018)
12. Yilmaz, H.B., Tugcu, T.: Location estimation-based radio environment map construction in fading channels. Wirel. Commun. Mob. Comput. **15**(3), 561–570 (2015)
13. Sun, G., Van de Beek, J.: Simple distributed interference source localization for radio environment mapping. In: 2010 IFIP Wireless Days, pp. 1–5. IEEE (2010)
14. Sato, K., Inage, K., Fujii, T.: Radio environment map construction with joint space-frequency interpolation. In: 2020 International Conference on Artificial Intelligence in Information and Communication (ICAIIC), pp. 051–054. IEEE (2020)

15. Isselmou, Y.O., Wackernagel, H., Tabbara, W., Wiart, J.: Geostatistical interpolation for mapping radio-electric exposure levels. In: 2006 First European Conference on Antennas and Propagation, pp. 1–6. IEEE (2006)
16. Song Wenjia, H.H., Tianyu., Z.: Spectrum sensing algorithm for UAV communication based on LSTM neural network. Sci. Technol. Wind **35**(12), 7 (2019)
17. Zhang Han, H.Y., Hang., J.: Construction method of electromagnetic spectrum map based on residual autoencoder. Radio Commun. Technol. **49**(2), 7 (2023)
18. Zhou, B., Zhang, Y., Chen, X., Shen, S.: Fuel: fast UAV exploration using incremental frontier structure and hierarchical planning. In: International Conference on Robotics and Automation (2021)
19. Debroy, S., Bhattacharjee, S., Chatterjee, M.: Spectrum map and its application in resource management in cognitive radio networks. IEEE Trans. Cogn. Commun. Networking **1**(4), 406–419 (2016)

Author Index

Q. Yu (Ed.): SINC 2023, CCIS 2057, pp. 165–166, 2024.
https://doi.org/10.1007/978-981-97-1568-8

Printed in the United States
by Baker & Taylor Publisher Services

Printed in the United States
by Baker & Taylor Publisher Services